Conservation chemistry – an introduction

Ted Lister and Janet Renshaw

Conservation chemistry – an intrdoduction

Written by Ted Lister and Janet Renshaw

Edited by Colin Osborne and Maria Pack

Designed by Imogen Bertin

Published and distributed by Royal Society of Chemistry

Printed by Royal Society of Chemistry

For further information on other educational activities undertaken by the Royal Society of Chemistry write to:

Education Department
Royal Society of Chemistry
Burlington house
Piccadilly
London W1J 0BA

Information on other Royal Society of Chemistry activities can be found on its websites:
http://www.rsc.org
http://www.chemsoc.org
http://www.chemsoc.org/LearnNet contains resources for teachers and students from around the world.

ISBN 0–85404–395–0

British Library Cataloguing in Publication Data.

A catalogue for this book is available from the British Library.

RS•C

Foreword

The chemical sciences and their applications are all around us. Many museums and galleries have scientific departments. Collectors of objects that may not even be 'antique' in the strict definition of the word often need to have some scientific knowledge to stop their collection deteriorating.

This resource shows how chemical techniques are used in conserving objects made from a wide variety of materials and seeks to introduce some of the ethical considerations of conservation to students. It is hoped that this will encourage teachers and students to consider the chemical sciences in their widest context and to reinforce the beneficial aspects of chemistry in unlikely contexts.

Dr Simon Campbell
President, Royal Society of Chemistry

RS•C

Using the resource on the Internet

Sample student worksheets can be downloaded from **http://www.chemsoc.org/networks/learnnet/conservation.htm**. They are available as photocopiable masters (as pdf files or as Word documents) that teachers may tailor it to their own requirements.

RS•C

Contents

Plastics conservation – Barbie™ and friends

RS•C

Introduction

This material has been compiled (with permission) by Ted Lister from the book by Anita Quye and Colin Williamson, *Plastics collecting and conserving*, Edinburgh: NMS Publishing, 1999.

Teachers may find the above book a useful source of further reading if required.

Acknowledgements

The Royal Society of Chemistry thanks the following people for their help in producing this resource:

Brenda Keneghan, Victoria and Albert Museum, London
Susan Mossman, The Science Museum, London
Anita Quye, National Museums of Scotland, Edinburgh
Colin Williamson, The Plastics Historical Society

The resource

Three pieces of student material are presented:

- *A brief history of plastics;*
- *The decay and degradation of plastics*; and
- *Conservation of plastics.*

Each passage is set in a context of the collection, care, identification and display of plastic objects in museums and by private collectors. Many people think of objects made of plastic as 'throwaway' and do not consider them as collectable items or ones that might be found in museums. In fact there are increasing numbers of plastic objects in museums as well as in private collections and many are increasing in value. To give just one example, some Barbie™ dolls can change hands for thousands of pounds. It is also a misconception that plastics do not decay easily – many of them do, and this raises issues about how best to preserve them.

The issues examined in the passages include the chemistry of plastics, some of their history, ways in which plastics degrade, identifying different types of plastics and ways to store and display them to minimise deterioration.

Each piece consists of reading material for post-16 students (although *A brief history of plastics* may also be suitable for able pre-16 students) interspersed with questions to fulfil the following functions:

- as summative comprehension questions, *ie* for students to read the material and answer the questions to test their understanding;
- as formative comprehension questions *ie* to help students' understanding as they read the chapter; and
- as a way of highlighting that the chemical principles used by chemists in real life situations are the same as those learned in post-16 study.

RS•C

Chemical nomenclature

The question of systematic names versus everyday names is particularly problematic with polymers, where everyday names are regularly used and recognised and systematic ones are often complex. In general, everyday names have been used in this publication with systematic ones in brackets where this is thought to be helpful, for example where the structure of the polymer is being discussed rather than the name being used simply to identify the material. In cases of doubt, we have erred on the side of simplicity and readability – Terylene is much more easily understood than poly(ethenediyl-1,4-benzenedicarboxylate)! Names that are registered trademarks, such as Terylene™, Nylon™ and Perspex™ are spelt with a capital letter. Other names such as polyvinyl chloride (abbreviation pvc) and polythene (abbreviation pe) begin with a lower case letter.

RS•C

Teachers' notes

Answers

A brief history of plastics

1. a) Many answers are possible including a variety of polymeric materials. Non-polymer based materials that fit the definition could include clay, metals, glass *etc*.

   b) Methods of shaping include carving, cutting, machining, forging *etc*.

2. a) The kink occurred in the mid-1970s.

   b) This follows the so-called Yom Kippur war between Israel and Egypt. Many oil-producing countries in the Middle East increased their crude oil prices in protest at the USA's support for Israel. This in turn increased the price of plastics (which are made from products obtained from crude oil) and thus reduced demand.

3. Bois durci means literally 'hardened wood'.

4. a) Vulcan was the Roman god of fire. The fire refers to the heat used to vulcanise rubber.

   b) Ebony is a hard, black wood so that Ebonite resembled ebony.

5. a) Elephants are relatively rare, they would have to be hunted, killed and the ivory imported. Only a relatively small amount of ivory is obtained from each elephant.

   b) Many people consider hunting cruel, and elephants are becoming increasingly rare.

   c) The ivory would have to be shaped into a perfect sphere.

6. Carbon dioxide (and carbon monoxide), water, nitrogen oxides.

7. The fibres have a much greater surface area exposed to oxygen in the air.

8. Many possible answers, for example saucepan handles, parts in car engines such as distributor caps.

9. A synthetic, thermoplastic condensation polymer.

The decay and degradation of plastics

1. a) 350 000 J mol^{-1}

   b) 350 000 / 6 x 10^{23} = 5.83 x 10^{-19} J per bond

   c) $\nu = 8.8 \times 10^{14}$ sec^{-1}

   d) This is at the borderline between the violet and ultraviolet regions of the spectrum.

2. Cross-links prevent polymer chains sliding past one another and prevent the material stretching and bending.

3. a) Reactions normally slow down as the concentrations of the reactants decrease.

   b) If one of the products catalyses the reaction, the rate of the reaction first decreases, as explained above. Then, as the catalyst appears, it increases and decreases again as the reactants are used up, see Figure 1.

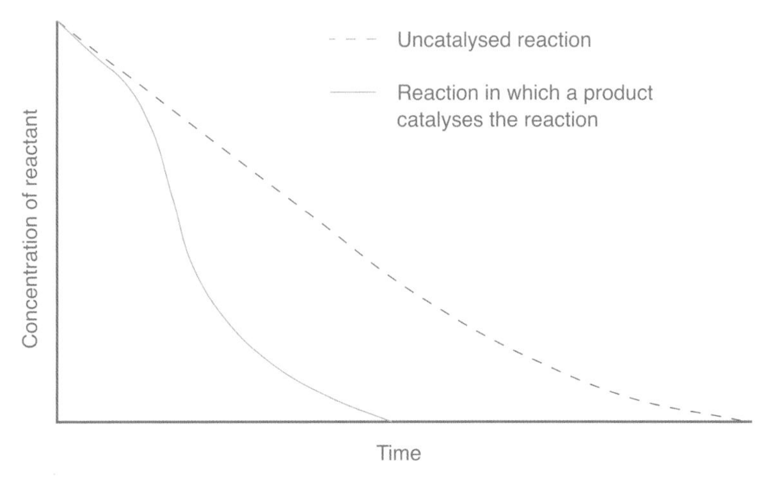

Figure 1

4. a)
   - 'a few days'; disposable contact lenses,
   - 'a few weeks'; carrier bag,
   - 'a few months'; chocolate bar wrapper
   - 'a few years'; washing-up bowl, toy, car bumpers
   - 'many years'; electrical cable insulation, hip joint

   Some variations are acceptable.

   b) A chocolate bar wrapper must last long enough to allow for the distribution time and shelf life of the bar. Ideally, it would then decay to harmless products but in practice it will present a disposal problem.

5. a) It is heated.

   b) This will speed up any reactions, such as chain breaking, which cause degradation.

6. Elimination.

7. A base.

8. a) Decolourisation of bromine solution.

   b) In a solid, only the double bonds at the surface are exposed to the bromine solution.

   c) See Figure 2.

$$-\mathrm{CH}=\mathrm{CH}- + \mathrm{Br_2} \longrightarrow -\mathrm{CHBr}-\mathrm{CHBr}-$$

Figure 2

9. Hydrogen chloride gas will be unable to escape, its concentration in the case will rise and its catalytic effect will increase.

RS•C

10. The sodium carbonate would react with the hydrochloric acid, thus slowing down the decay of the pvc.

$Na_2CO_3(s) + 2HCl(g) \rightarrow 2NaCl(s) + H_2O(l) + CO_2(g)$

11. See Figure 3.

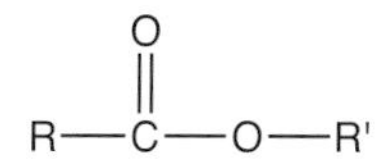

Figure 3

12. a) See Figure 4.

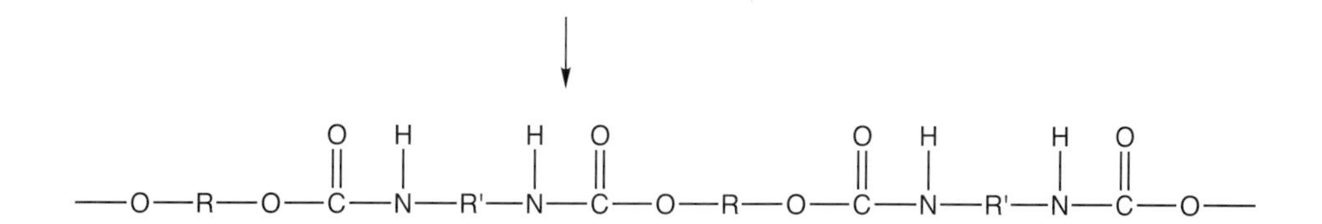

Figure 4

b) A reactive site on each end of a chain is needed to make a polymer so that as each monomer adds onto the growing chain, further reaction is still possible.

c) A triol allows a branched chain to form and thus leads to the possibility of cross-linking between chains.

d) Cross-linking is likely to make the polymer more rigid.

13. a)

b) See Figure 5.

~120°

Figure 5

14. Foams have a large surface area exposed to oxygen in the air. Paint protects the polyurethane from oxygen in the air.

15. Water.

16. $ZnO + 2HNO_3 \rightarrow Zn(NO_3)_2 + H_2O$

17. A smell of vinegar. This is the ethanoic acid released by the decaying polymer.

18. a) A strong acid, HA, dissociates completely into H^+ and A^- in solution. A weak acid dissociates only partly and an equilibrium exists the solution which contains mostly undissociated HA.

b) Ethanoyl chloride.

RS•C

Conservation of plastics

1. Displaying an object means that it is exposed to a light and to a variety of chemicals – oxygen and other gases in the air and water vapour for example. It will also be exposed to temperature changes and may be touched.
2. Look for sensible suggestions. The hinges in the joints may be made of metal. Hair could be made of Nylon and eyes of glass. Clothing could be made from a variety of materials both synthetic and natural, such as cotton.
3. a) 1.08 g cm^{-3}.

   b) Polystyrene.

   c) There may be other types of plastic with this density.

   Many plastics contain additives such as fillers that affect the density.

   Accuracy of measurement – a small inaccuracy in the measurement of mass or volume would result in a different value for density. Since many plastics have similar densities, this could result in a misidentification.

   d) Place the piece in a measuring cylinder part-full of water. Note the volume increase.

   e) The density will be much less than for the non-foamed plastic – see the Table in the question.
4. a) Nitric acid. Ethanoic acid (acetic acid).

   b) Both these plastics are long chain hydrocarbons. Wax candles and paraffin are made from shorter chain hydrocarbons that could be produced by breaking the longer chains.

   c) Carbon dioxide. It is acidic.

   d) Carbon monoxide.
5. a) See Figure 6.

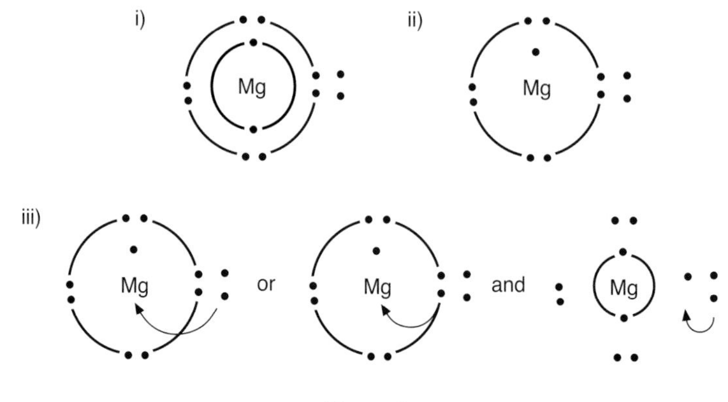

Figure 6

   b) Two: (i) directly from shell 3 to shell 1 and (ii) indirectly – from shell 2 to shell 1 and then from shell 3 to shell 2. There are therefore three frequencies in the spectrum.
6. This is a difference of 20 °C and would reduce reaction rates fourfold, thus quadrupling the life of the object.
7. a) 1.66×10^{-9}.

RS•C

b) 6.6×10^{-10}.

c) The first is approximately 2.5 times larger than the second.

d) This supports the rule (if only as a rule of thumb) because the rule would predict the first to be twice as big as the second.

8. a) The increased surface area compared with a solid lump will increase the rate of reaction of the iron with the oxygen.

b) iron + oxygen → iron(III) oxide

$4Fe(s) + 3O_2(g) \rightarrow 2Fe_2O_3(s)$

c) Volume of case = 1 x 0.5 x 0.5 = 0.25 m^3 = 250 dm^3
Air is 20% oxygen, so the case contains 50 dm^3 of oxygen
From the equation, 3 moles (3 x 24 = 72 dm^3 at room temperature) of oxygen react with 4 moles (4 x 56 = 224 g) iron.
So 155.6 g of iron is required.

The main assumption is that at room temperature the volume of 1 mole of gas is 24 dm^3. Other assumptions are that the case has no leaks and that the volume of contents of the case can be ignored.

Practical work

The following experiment is about identifying plastics according to their density. An alternative version of this experiment is given in K. Hutchings, *Classic Chemistry Experiments*, London: Royal Society of Chemistry, 2000, pp27–32. The student sheet is available at http://www.chemsoc.org/networks/learnnet/classic_exp.htm (look at files 11-20).

Experiment – Identifying plastics according to their densities
This experiment could be done as a class experiment or a demonstration.

Materials will float in liquids that have a greater density than they do. For example, polythene (density about 0.9 g cm^{-3}) will float in water (density 1.0 g cm^{-3}) but pvc (density about 1.2 g cm^{-3}) will not.

Solutions of salt (sodium chloride) have greater densities than pure water. Pure water has a density of 1.0 g cm^{-3} while a saturated salt solution has a density of about 1.3 g cm^{-3}.

One way of comparing the densities of plastics is to see whether or not they float in salt solutions. Take small pieces of the following plastics: polythene, polyester, pvc, phenolic, Perspex, polystyrene. Place them in a 250 cm^3 beaker and almost fill the beaker with tap water to which one drop of washing-up liquid has been added. (The washing-up liquid is to prevent bubbles sticking to the plastic pieces. These would tend to make them float.) You should find that the polythene floats, because its density is less than that of water. Now add salt, a large spatula full at a time. Stir after each addition to dissolve the salt. Continue until you have added about 70 g of salt. You should notice that the various plastics begin to float one by one as the density of the salt solution increases. In which order do the plastics start to float? What does this tell you about the densities of these plastics?

Small samples of the required plastic materials can be obtained as follows:

- polythene – bread bags, bin liners, carrier bags, food bags (either high or low density polythene is suitable for this experiment);

- polyester – fizzy drinks bottles (the transparent body of the bottle, not the black base);
- pvc – cooking oil or shampoo bottles;
- phenolic – electrical fitting such as plug sockets (normally dark-coloured);
- Perspex™ – Perspex™ sheet sold in DIY shops, car rear light lenses; and
- polystyrene – disposable cups from drinks machines (not expanded ones) translucent plastic egg boxes (not expanded ones).

Some further suggestions for sourcing plastic samples are given in K. Hutchings, *Classic Chemistry Experiments,* London: Royal Society of Chemistry, 2000, p28, and *Nuffield Co-ordinated Sciences, Teachers' Guide.* Harlow: Longmans, 1988, p214. The former also includes a table of densities of various plastics.

RS•C

A brief history of plastics

The word 'plastic' means literally 'able to be moulded'. A more modern definition is 'a material that can be moulded or shaped into different forms under pressure and/or heat'. By these definitions, clay, for example, is plastic. Nowadays when we use the word plastic we usually mean one of a group of synthetic (man-made) materials such as polythene, pvc, Bakelite™, Nylon™ *etc.* All these materials are polymers – large molecules made up of many smaller molecules (called monomers) chemically bonded together. These materials are also plastic in the original sense in that they can be moulded into shape (although some of them require additives to make them easily mouldable). This ability to be moulded is one of the main advantages of plastic materials. Some types can be moulded only once and then retain their shape until they are destroyed, others can be remoulded several times.

Q1. **a)** Think of as many materials as you can which fit the definition of a plastic: 'a material which can be moulded or shaped into different forms under pressure and/or heat'. Which ones on your list are synthetic polymers?

b) What other methods are there for shaping materials other than moulding them into shape?

Many synthetic polymers are of relatively recent origin – most of them have been developed in the last 50 years and none is much more than 150 years old. Compare this with metals, which have been used by mankind for over 5000 years and materials such as wood and stone, which have been used for over 2 million years.

The use of plastic materials has increased enormously in the last 50 years (Figure 1).

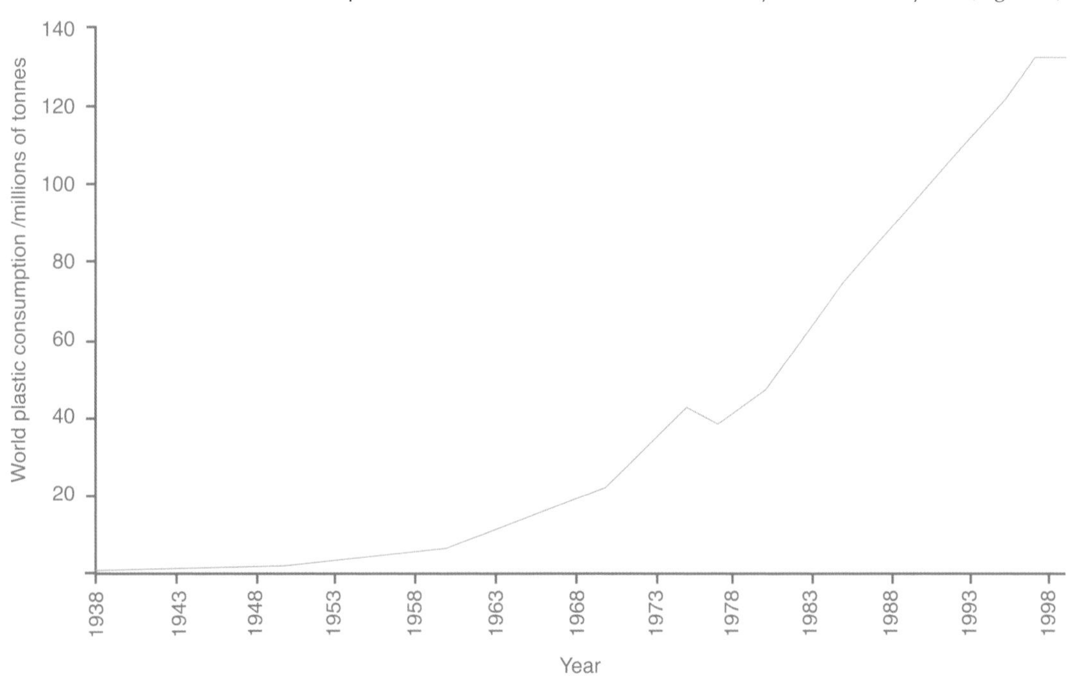

Figure 1 World plastic consumption during the 20th century

Q2. Look at the graph in Figure 1. There is a significant kink in plastics production.

a) Use the graph to work out the approximate date of this kink.

b) A world event around this date caused this sudden drop in plastics production. Find out what this was.

RS•C

Sources of polymers

Polymers themselves are not new. Cellulose is a polymer made from sugar monomers. It is found in plants and has been around as long as the plants themselves. Rubber is another polymer from a plant source, the rubber tree, Figure 2. This time the monomer is a molecule called isoprene (methylbuta-1,3-diene). The formulae of cellulose and of rubber are shown in Figure 3.

Figure 2 Natural rubber is the sap of the rubber plant

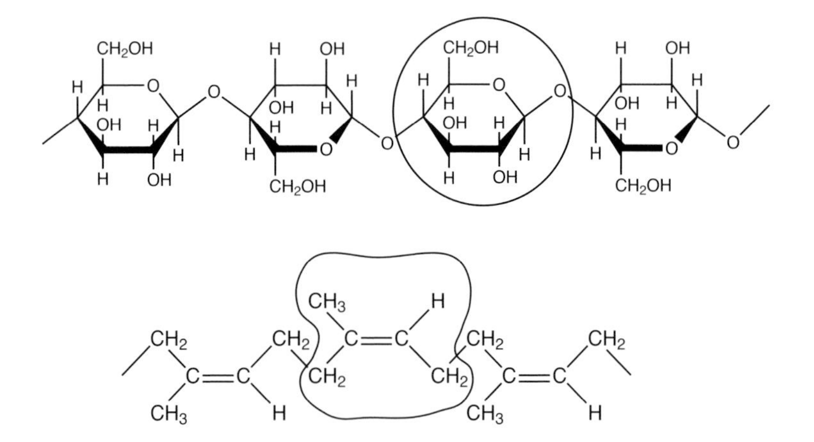

Figure 3 Parts of the polymer molecules of cellulose (top) and rubber (bottom) with the monomer units ringed

Rubber can be moulded and is therefore a plastic; it can be used as it is, although it is usually treated chemically to make it harder. Cellulose, on the other hand, cannot be moulded as it is. It can, however, be treated chemically to convert it to a form which can be moulded. Treating cellulose, obtained from cotton, with nitric and sulfuric acids converts the cellulose into a form that can be moulded into shape. This new form is an example of a modified (or semi-synthetic) polymer. This polymer is called cellulose nitrate by chemists and has the trade name Celluloid™.

So there are three types of polymer:

- natural polymers such as rubber, which can be used with little or no chemical treatment;
- natural polymers which need chemical treatment to make them useful – these are called modified (or semi-synthetic) polymers; and
- synthetic (man-made) polymers that are made from raw materials such as products from crude oil which are not themselves polymers.

RS•C

Natural polymers

Not surprisingly, the first polymers to be used by mankind were natural polymers. Some examples include:

- animal horns and hooves – these can be softened in boiling water so that they can be shaped;
- beeswax – taken from beehives – this can be softened or melted by heating, and hardens on cooling;
- shellac – this is a substance, given out by some insects, that can be melted and moulded; and
- bois durci – the protein albumen, found in blood and egg white, mixed with powdered wood and made into a mouldable polymer.

Some of these polymers have been used for up to 2000 years. Some examples are shown in Figure 4.

Figure 4 Clockwise from top left: horn brooches, shellac cases, Bois Durci plaque of the composer Haydn, beeswax candles
(Photographs courtesy of Smile Plastics Ltd and Photodisc.)

Q3. Bois durci was developed by a Frenchman, Francois Charles Lepage. Use a French dictionary to help you work out what 'bois durci' means. Hint: the word 'durci' comes from the verb durcir, which is linked to the word dur.

Modified polymers

As chemical understanding and techniques developed, it became possible to use chemistry to change the properties of natural polymers to tailor them to particular uses. One of the first examples was the vulcanisation of rubber which was discovered in about 1840 by Thomas Hancock in the UK and Charles Goodyear (who gave his name to the Goodyear tyre company) in the USA. Natural rubber is soft and sticky. Hancock and Goodyear found that heating it with sulfur made rubber much harder, but that it still retained its flexibility. The new material was much more useful for making articles such as soles for shoes. The more sulfur that was added, the harder the rubber became.

Adding 30% sulfur produced a hard, black material called 'Ebonite' or 'Vulcanite' which was used for making items such as combs, brooches and parts of pens. It was also used as an electrical insulator. Some examples are shown in Figure 5.

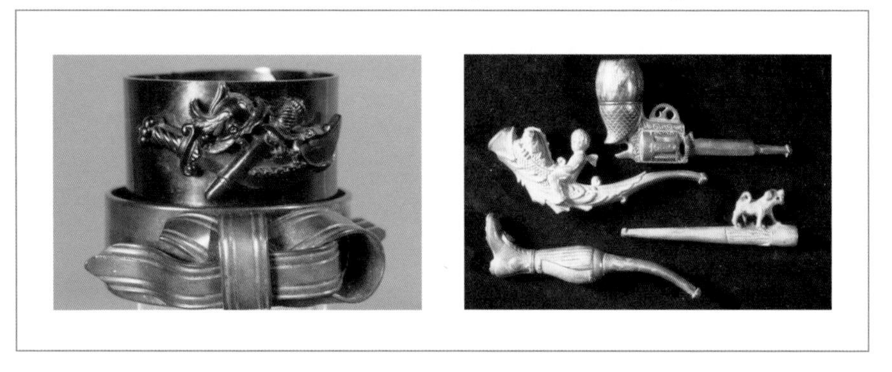

Figure 5 Vulcanite bracelets, vulcanite pipes. Vulcanite is also known as ebonite

Q4. **a)** Find out about the Roman god Vulcan. Suggest why the above process is called 'Vulcanisation'.

b) Find out about ebony, which is a type of wood. Suggest why the modified rubber polymer was sometimes called Ebonite.

Hancock and Goodyear's discoveries were made largely by trial and error. We now know that natural rubber is made up of long chain molecules rather like strands of cooked spaghetti. When natural rubber is stretched, these molecules can slide past one another and when the stretching stops, they stay in their new positions. Reaction with sulfur forms chemical links (called cross-links) between the chains so that they can no longer slide past each other, see Figure 6. If there are just a few of these links, the material tends to pull back to its original shape when stretched, making it elastic. More links make it impossible to stretch the material at all, and it is hard. So, depending on the number of cross-links, rubber can vary from a consistency like chewing gum, to an elastic material like a rubber band, to a hard and rigid material.

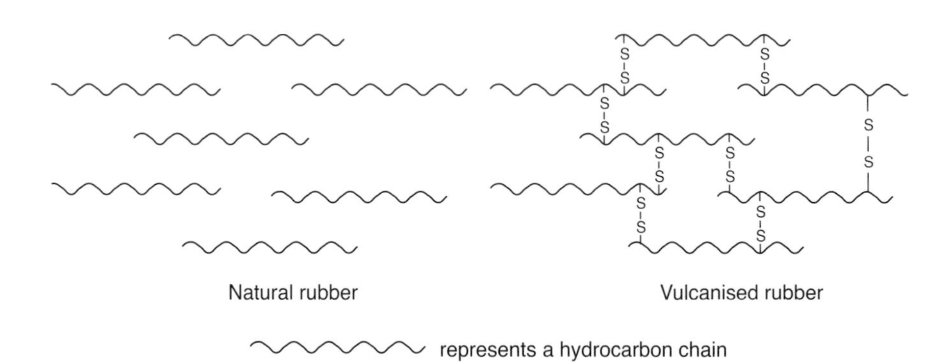

Figure 6 The vulcanisation of rubber

In the 1840s, the idea that substances were made up of small particles called atoms and molecules was new and few scientists really accepted it. Certainly the idea of long chain molecules was not accepted until about 100 years later following the work of a German chemist called Hermann Staudinger.

RS•C

Great balls of fire!

Before synthetic plastics appeared on the scene, the only materials available were natural ones. This meant that quite ordinary objects had to be made from materials that we would now consider to be very expensive and inappropriate. For example billiard balls had to be made from ivory, one source of which is the tusks of elephants. Chemists then began to treat natural materials with chemicals to improve their properties. An example of this was the treatment of cellulose with nitric and sulfuric acids to form cellulose nitrate (a semi-synthetic polymer). One of the first uses of cellulose nitrate plastics was to make billiard balls.

Q5. a) Why would billiard balls made from ivory be expensive?

b) Why would many people nowadays consider that it would be inappropriate to make billiard balls from elephant ivory?

c) What manufacturing problem would have to be overcome when making billiard balls from ivory?

One problem with cellulose nitrate is that it is unstable and flammable. In fact chemically it is very similar to the explosive, guncotton. So it presented a significant fire risk. It has even been suggested that the energy of a collision between two balls on the table could set off a minor explosion. This has led to a story (probably apocryphal) dating from the late 1800s about a letter to a manufacturer of billiard balls from a saloon owner in Colorado, USA. This is supposed to have said that the saloon owner did not mind the explosions so much but that as soon as one occurred, every man in the saloon instantly pulled a gun! This, of course was the era of the American "wild west".

Fibres could also be made from cellulose nitrate. They were used as a substitute for silk called Chardonnet silk, after its inventor, Hilaire de Chardonnet. This material also suffered from flammability problems; a joke from the 1890s (which would nowadays be considered politically incorrect) suggested that the ideal present for a mother-in-law was a dress made from Chardonnet silk, and a match.

Q6. The elements that make up cellulose nitrate are carbon, hydrogen, oxygen and nitrogen. Suggest the products formed when cellulose nitrate burns in air.

Q7. Cellulose nitrate fibres can be ignited by touching them with a lighted match but billiard balls made of the same material cannot. Explain this.

Because cellulose nitrate is so flammable, chemists searched for a substitute. One approach was to react cellulose with other acids. One acid that (indirectly) produced a useful plastic was acetic acid (now called ethanoic acid). The resulting polymer is called cellulose acetate (cellulose ethanoate) and is still used in Rayon fabrics.

RS•C

Synthetic polymers

The first completely synthetic polymer was Bakelite™, named after its inventor, the Belgian-born American chemist Leo Baekeland who first produced it in 1909. He made it by reacting two chemicals together, phenol (then obtained from coal) and methanal (then obtained from wood). These two chemicals both consist of small molecules that become linked together in the reaction. So Baekeland had actually made a polymer from scratch rather than changing one that had already been made naturally. In the case of Bakelite™ the molecules were linked not so much in chains, as in a network – chains held together by links between them. This meant that Bakelite™ could be moulded into shape before the chemical reaction started but, once made, there were so many links between the chains that it could not be softened by heat and re-moulded. This type of plastic, which can be moulded only once (when it is first made) and cannot then be re-moulded, is described as thermosetting.

Q8. Suggest some everyday uses that require a material that does not soften when it is heated.

Bakelite™ was used for a variety of objects such as ashtrays, the cases of radios and even coffins, see Figure 8. Similar polymers are still used for making electrical sockets, because they are good electrical insulators.

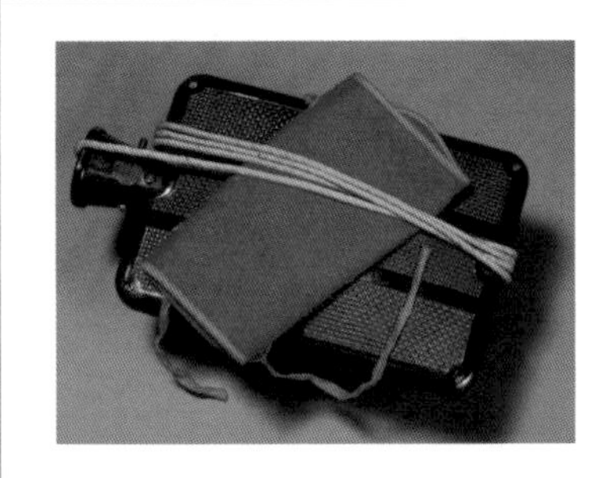

Figure 8 Objects made from Bakelite™

Synthetic thermoplastics

In the 1930s, some new types of synthetic plastic began to be made which had distinctly different properties to those of Bakelite and other thermosetting plastics. They soften (and can be moulded) when heated, and harden when cooled, a cycle which can be repeated over and over again. These plastics are described as thermosoftening or thermoplastic, which literally means 'heat mouldable'.

The first of these was polyvinyl chloride now called poly(chloroethene) by chemists but normally known by it initials, pvc. This polymer is made from the monomer vinyl chloride (chloroethene). The formula of chloroethene is C_2H_3Cl or $CH_2{=}CHCl$. The = sign means that there is a double bond between the two carbon atoms. This monomer can be made to link together in long chains, Figure 9.

RS•C

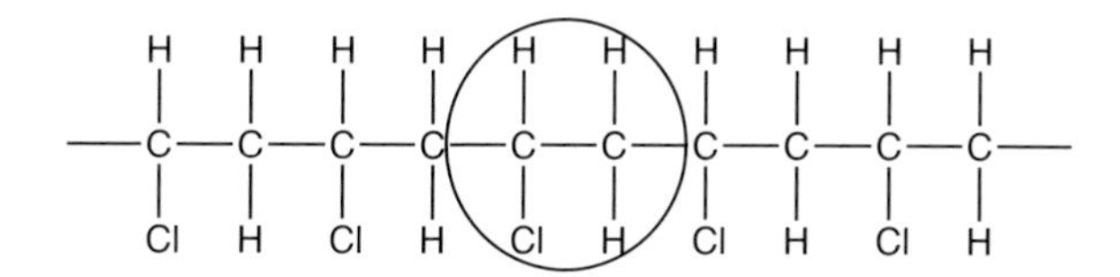

Figure 9 Part of the pvc molecule with the monomer unit ringed

The simplest formula (called the empirical formula) of the polymer is the same as that of the monomer – C_2H_3Cl. Nothing has been added or taken away, the monomers have just linked together. This type of polymer is called an addition polymer – the name stressing that the monomers have just added together.

There are now many addition polymers of this type known and used. Their monomers all have a carbon-carbon double bond and a general formula of CH_2=CHX, where X can stand for a variety of atoms or groups of atoms. For example in polythene (poly(ethene)), X is a H atom, in polypropylene (poly(propene)), X is a CH_3 group and in polystyrene (poly(phenylethene)), X is a ring of six carbon atoms, five of them with hydrogens attached.

A few years later, a different type of thermoplastic was developed. One of the first examples was the material now known as Nylon™, developed for the Du Pont company in the USA by a team led by the chemist Wallace Carothers. Nylon™, and other polymers of the same type are made from two different monomers, which we can call H-X-H and HO-Y-OH where X and Y represent chains of carbon and hydrogen atoms. When the monomers react together, an H from H–X–H and an HO from HO–Y–OH react together to form a molecule of water, HOH or H_2O, and X and Y are joined with a covalent bond, Figure 10.

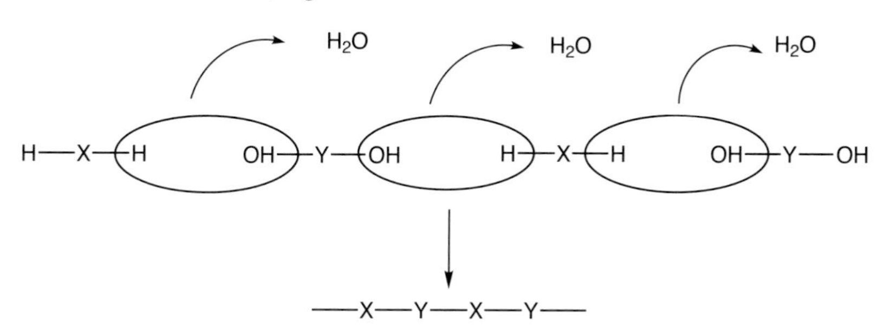

Figure 10 Making Nylon™

The technical name for two (or more) molecules joining together and eliminating a small molecule as they do so is 'condensation' so polymers like this are called condensation polymers. The molecule that is eliminated is often water but may be another small molecule such as hydrogen chloride (HCl).

RS•C

What's in a name?

Nylon is a trade name rather than a proper chemical name. There have been many stories about the origin of the name – one of the most common being that it was derived from New York London, the two headquarters of the Du Pont company. The truth seems to be that it was simply a word dreamed up by Du Pont's advertising department and that the polymer could just as easily have been called Nulon. The word Nylon was thought to trip off the tongue easily and sounded a little like Rayon, which is the trade name for cellulose acetate and was also used to make clothing fibres.

Systematic chemical names are rarely used for plastics. This is because they are often too long for everyday use and too cumbersome to trip off the tongue. For example the systematic name of Terylene™ is poly(ethenediyl-1,4-benzenedicarboxylate)! Incidentally, although Terylene™ is a trade name, it does have a chemical history. The non-systematic name of one of the monomers from which Terylene™ is made is terephthalic acid (benzene-1,4-dicarboxyloic acid).

Nylon™ was developed to replace silk for making stockings

Summary

We have seen three ways of classifying plastics:

1. Natural, modified (semi-synthetic) or synthetic describes the origin of the polymer.
2. Thermoplastic (thermosoftening) or thermosetting describes how the plastic reacts to heat.
3. Addition or condensation describes the type of chemical reaction used to make the polymer.

All three classifications can be used to describe a single material. For example rubber is a natural, thermoplastic, addition polymer. Bakelite is a synthetic, thermosetting, condensation polymer.

Q9. How would you describe Nylon™ in the same way that rubber and Bakelite are described above?

The decay and degradation of plastics

Although people often think that they last forever, plastics do decay over a period of time as chemical changes occur which affect their structures. Different plastics decay in different ways and at different rates, so it difficult to talk about the decay of plastics in general. However, most plastics have in common the fact that they are organic polymers based on long chains of carbon atoms linked by covalent bonds. There may be cross-links between these chains and there may also be side groups attached to the chains.

The factors that lead to decay of polymers are the same for most types of plastic. They include:

- light;
- oxygen from the atmosphere;
- moisture from the atmosphere;
- additives, such as fillers or plasticisers, from within the plastic itself; and
- mechanical stress, such as being stored under pressure.

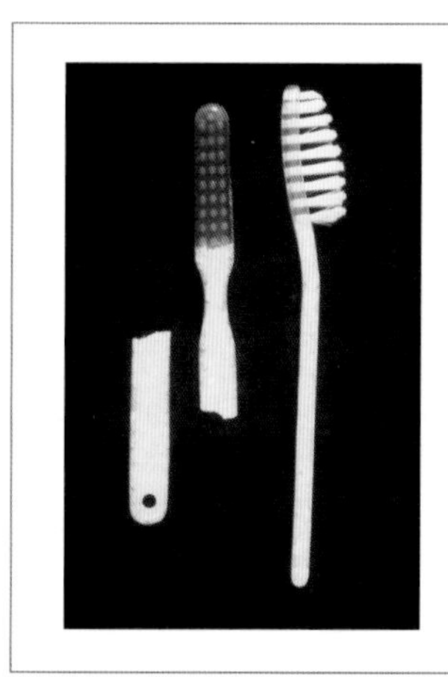
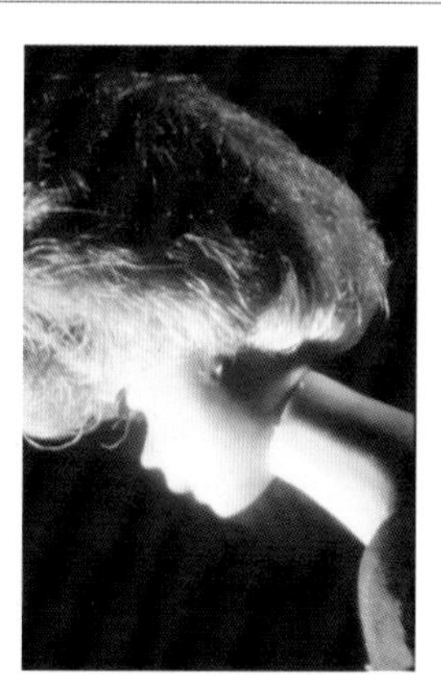
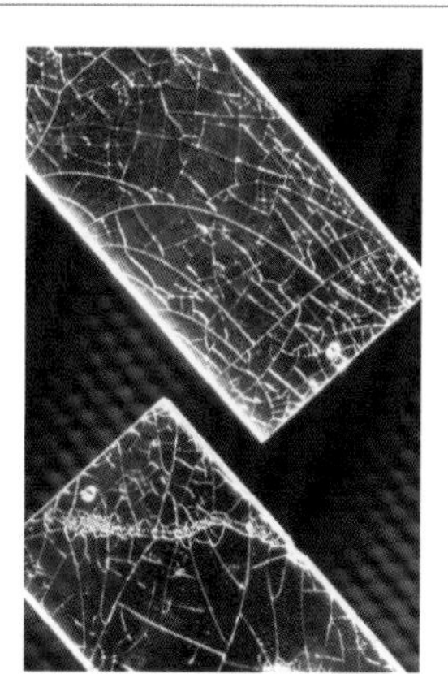

Figure 1 Decay of plastics: cellulose nitrate tooth brushes, a Barbie™ doll, Perspex labels, cellulose nitrate set square
(Pictures courtesy of Smile Plastics and the National Museum of Scotland.)

These lead to a number of general types of chemical change within the polymer molecules.

1. Shortening of the polymer chains
   If bonds within the polymer chains are broken, a polymer with shorter chains can be formed. This will normally have properties that are less useful than those of the original polymer – solid polymers may start to crumble, for example.
2. Cross-linking of the polymer chains
   Chemical changes that form new bonds between the existing polymer chains will normally lead to the polymer becoming less flexible and more brittle.

3. Chemical changes to the side groups in polymers
   These can often lead to the release of small molecules such as water or acids. The loss of these changes the structure of the polymer. The released molecules themselves can often bring about further changes or, in some cases, catalyse the reaction that produced them in the first place.

The effect of light on polymers

Light can cause polymer chains to shorten (this involves bond-breaking) and can also make cross-links form between chains (this involves bond-making). You may find it difficult at first sight to see how the same factor can cause both these changes.

When light energy breaks a carbon-carbon bond, the bond breaks homolytically, that is, one of the two shared electrons in the bond goes to one of the carbon atoms and the other goes to the other carbon atom, so that each has an unpaired electron (shown in equations by •). This is because both carbons have exactly the same electronegativity (electron-attracting power). If a carbon-hydrogen bond is broken, the same thing happens because hydrogen and carbon have almost the same electronegativity. The two situations are shown in Figure 2.

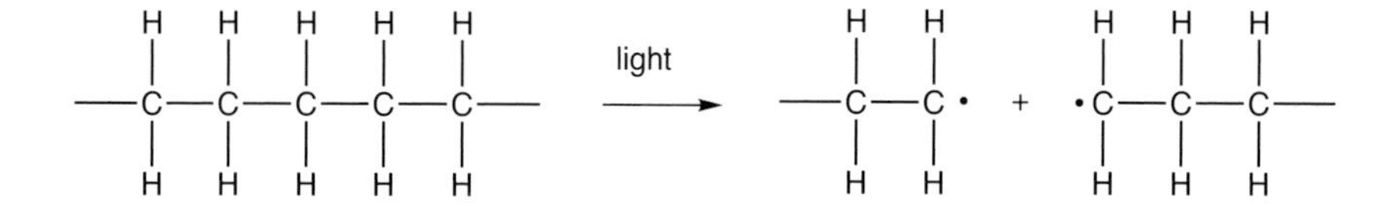

Figure 2 Bond-breaking caused by light

Species with unpaired electrons are highly reactive and are called radicals. They will react with almost anything to form a more stable species with paired electrons. Two radicals formed by breaking C-H bonds may come together and form a covalent carbon-carbon bond. This forms a cross-link between the chains, Figure 3. The radicals formed by the breaking of a C-C bond may react with some other species, possibly a hydrogen atom from elsewhere in the polymer. This will lead to two shorter chains, Figure 4.

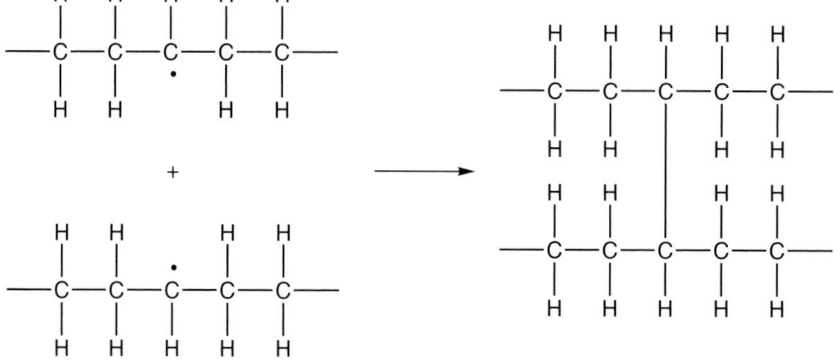

Figure 3 Two radicals may form a cross-link

Step 1

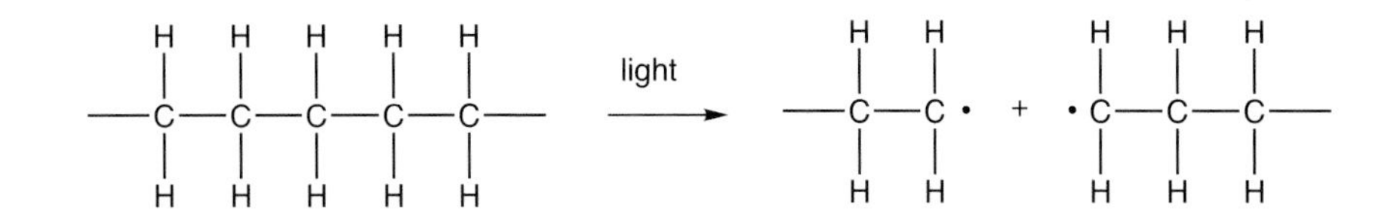

Step 2

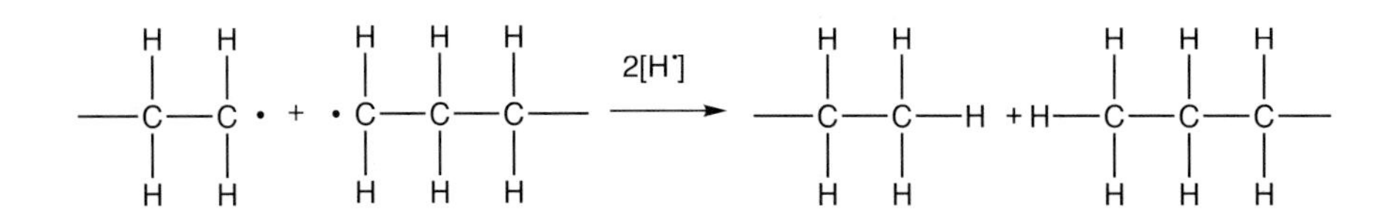

Figure 4 Chain shortening

Q1. Does visible light have enough energy to break chemical bonds? This question shows you how to find out. The bond energy of carbon-carbon single bonds is about 350 kJ mol^{-1}. A quantum of light energy (a photon) has an amount of energy, E, given by the equation E = hv, where h is Planck's constant, 6.6×10^{-34} Js, and v is the frequency of the light.

a) What is the energy of C-C bonds in J mol^{-1}?

b) What is this energy in J per bond? The Avogadro constant (the number of molecules in a mole) is 6×10^{23}.

c) Use E = hv, to find the frequency of light that has just sufficient energy to break a carbon-carbon bond.

d) Use a reference book or database to find in which region of the electromagnetic spectrum this light lies.

Q2. Explain why cross-links between polymer chains will lead to the material becoming less flexible. A diagram might help.

Q3. **a)** What happens to the rate of most chemical reactions as time goes on? Explain your answer.

b) What happens to the rate of a reaction in which one of the products is a catalyst for the reaction? Explain your answer. You may find that it helps to sketch graphs of concentration of reactant or product against time as part of your answer.

It is possible to add materials called stabilisers to some plastics to slow down their decay. However, many objects made of plastic are not designed to last for long and so it is often not cost-effective for manufacturers to use these.

Q4. Plastics can be used for:

A washing up bowl, disposable contact lenses, a carrier bag, the insulation on electric cables used for house wiring, a chocolate bar wrapper, a toy, a plastic hip joint, a car bumper.

a) For each of the plastic items in the list, estimate how long they are designed to last. Select from:

- 'a few days';
- 'a few weeks';
- 'a few months';
- 'a few years'; and
- 'many years'.

b) Explain your answer for the chocolate bar wrapper by describing what would happen if the material did not last long enough, and what would happen if it lasted too long.

We will look in a little more detail at how four common plastics decay.

Polyvinyl chloride, pvc (poly(chloroethene))

Like many plastics, pvc materials are made of more than just the polymer. The basic polymer is shown in Figure 5, but commercial materials have a number of additives.

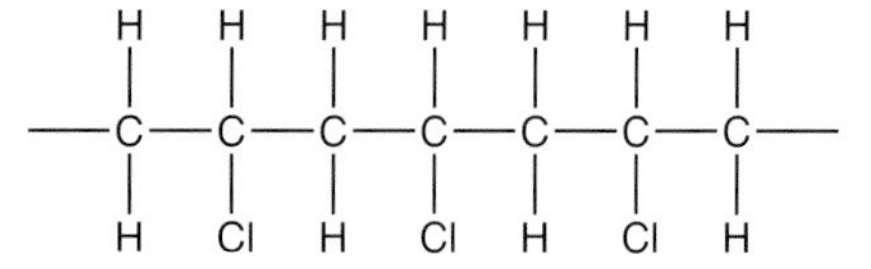

Figure 5 Part of a pvc molecule

These include plasticisers to increase the flexibility and stabilisers to help prevent the polymer degrading when it is moulded during manufacture and later. Plasticisers are small molecules that lie between the long chain polymer molecules and act like a lubricant, making it easier for the large molecules to slide past one another. Pvc plastics can be made with a wide range of flexibility. This is shown by the wide variety of objects that are made from pvc. This includes drainpipes, shoes, 'vinyl' records, hospital blood bags and hosepipes.

Stabilisers work in a variety of ways – some of them are molecules that react with the free radicals formed when ultraviolet light causes bond-breaking and prevent them causing further reaction.

Q5. **a)** What is done to pvc during manufacture to make it soft enough to be moulded?

b) Why is this treatment likely to make it degrade?

The main long-term degradation reaction of pvc is the loss of hydrogen chloride, HCl. This is an acidic gas that dissolves in water to form hydrochloric acid. The loss of HCl can be brought about by ultraviolet light and/or high temperatures. It results in the formation of carbon-carbon double bonds in the polymer chain – one for each molecule of HCl lost, Figure 6.

RS•C

```
     H   H   H   H   H   H   H
     |   |   |   |   |   |   |
  ---C---C---C---C---C---C---C---
     |   |   |   |   |   |   |
     Cl  H   Cl  H   Cl  H   Cl

             |  - HCl
             v

     H   H   H   H   H   H   H
     |   |   |   |   |   |   |
  ---C===C---C---C---C---C---C---
             |   |   |   |   |
             Cl  H   Cl  H   Cl

             |  - HCl
             v

     H   H   H   H   H   H   H
     |   |   |   |   |   |   |
  ---C===C---C===C---C---C---C---
                     |   |   |
                     Cl  H   Cl
```

Figure 6 Loss of hydrogen chloride from pvc

Q6. Classify the reaction by which pvc degrades as one of the following types: substitution, addition, oxidation, elimination, reduction.

The acidic hydrogen chloride acts as a catalyst for the degradation of pvc, so that once started it will tend to speed up. The polymer resulting from this reaction may have an arrangement of alternating double and single bonds, Figure 6. This is described as a conjugated system. Conjugated systems tend to absorb visible light, *ie* they are coloured. So one effect of this type of degradation is to make the polymer yellow and eventually darken in colour. The doll shown in Figure 7 shows the effect clearly. The head and hands of the doll, which have been exposed to light, have darkened while the body and legs, normally covered by clothes, have not.

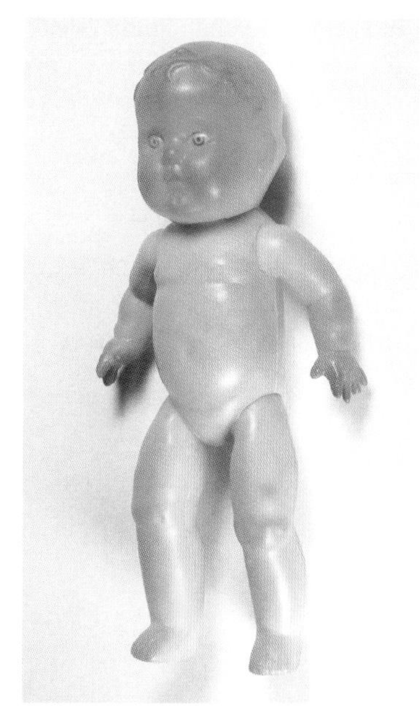

Figure 7 The exposed parts of this doll have darkened in colour

Q7. Since hydrochloric acid catalyses the degradation of pvc, suggest what sort of substance might be added to pvc to act as a stabiliser.

Did you know? The molecule β-carotene, Figure 8, is responsible for the orange colour of carrots and the pink of flamingos' feathers. It has a conjugated system of 11 alternating double and single bonds. It absorbs blue light and therefore reflects the red and yellow, so it looks orange.

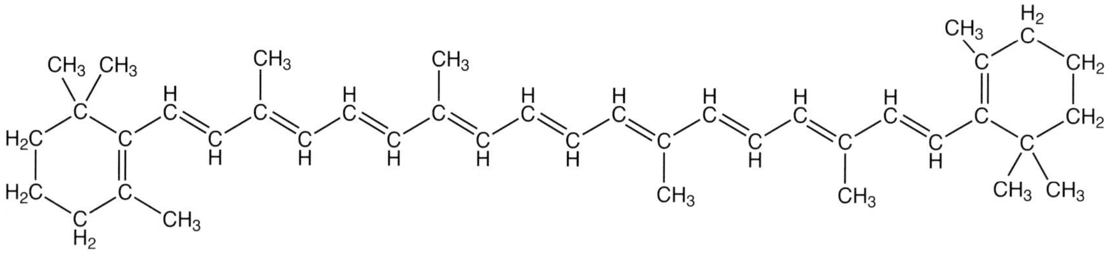

Figure 8 β-carotene

Q8. a) Give a simple chemical test for a compound containing carbon-carbon double bonds. This test is usually carried out on a solution of the compound under test.

b) Why might it be more difficult to carry out on a solid sample of plastic?

c) Write an equation to explain the test and its result.

Q9. Why might a pvc doll that was already degrading deteriorate more quickly if kept in a sealed case than if stored in a well-ventilated room?

Q10. A museum conservator wished to display a pvc object in a glass case to prevent damage caused by the public handling it. She suggested that placing a small dish containing sodium carbonate in the case might help to prevent degradation of the plastic. Explain why this might help and give any chemical equation(s) relevant to the protection process.

Another way in which pvc objects degrade is by loss of plasticiser molecules. These molecules are often liquid esters. It is sometimes possible to see beads of plasticiser on the surface of pvc and this is called 'weeping'. Loss of plasticiser may make the plastic shrink and become more rigid, both of which may cause it to crack.

Q11. Draw the functional group of an ester.

Polyurethanes

The polyurethanes are another versatile group of polymers. They contain the urethane group –NHCOO– that is formed by the reaction of an isocyanate (an –NCO group) with an alcohol. See Figure 9.

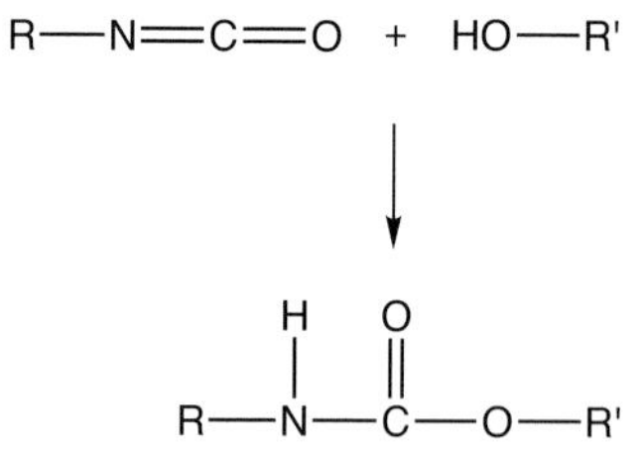

$$R{-}N{=}C{=}O + HO{-}R' \longrightarrow R{-}NH{-}C({=}O){-}O{-}R'$$

Figure 9 Formation of a urethane link

Q12. To form a polyurethane, two monomers are used, a diol and a di-isocyanate.

a) Give an equation for this reaction using HO–R–OH to represent the diol and O=C=N–R'-N=C=O to represent the diisocyanate and showing at least two molecules of each monomer.

b) Explain why a diol and a diisocyanate are needed to make a polymer.

c) Suggest what might happen if a small amount of a triol were added to the reaction mixture.

d) What effect might this have on the property of the resulting polymer?

Q13. Isocyanic acid, H–N=C=O is the 'parent' of the isocyanates. Draw a dot-cross diagram to show the bonding in this molecule and predict its shape.

Polyurethanes can be made to have a wide variety of properties making them suitable for many uses – from foams for cavity wall insulation and furniture fillings to fibres such as Lycra used in swimwear. The differences are partly explained by differences in chain lengths of the R groups in the monomers.

Foams, such as the padding behind the dashboards of cars and the filling of cycling helmets are made by introducing a gas into the liquid polymer during manufacture before it sets. This produces a honeycomb-like structure of tiny bubbles looking rather like the inside of a Crunchie® bar.

Polyurethanes degrade by oxidation, which causes them to discolour and then weaken. This is a particular problem with polyurethane foams and less of a problem with polyurethane items that have been painted.

Q14. Explain why the information in the sentence above is consistent with the decay of polyurethane being caused by oxygen in the air.

Cellulose nitrate

This is one of the oldest synthetic plastics. In fact it is only partly synthetic. It is made by reacting a natural polymer, cellulose (a poly-sugar or polysaccharide found in wood and cotton) with nitric and sulfuric acids. Some of the –OH groups on the cellulose molecule are converted to nitro groups (–ONO_2), Figure 10. Cellulose nitrate is the plastic from which old movie films were made and its degradation is a cause of serious concern for film historians as original copies of classic films become unviewable.

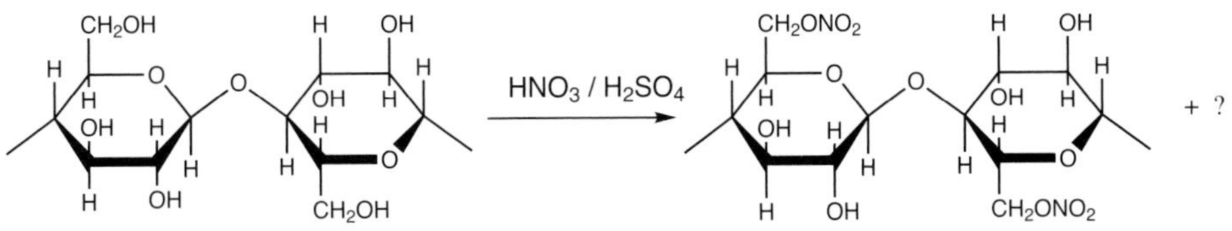

CH_2OH H OH H O H OH H H O H OH H H HNO_3 / H_2SO_4 CH_2ONO_2 H OH ... + ?

Figure 10 Formation of cellulose nitrate

Q15. Complete the equation above for the nitration of cellulose by working out what small molecule is eliminated, *ie* what does the question mark in the equation represent?

Changing the number of –OH groups per sugar molecule that are nitrated changes the properties of the final plastic.

Did you know? The nitration of cellulose has to be carefully controlled because the percentage of the –OH groups that are nitrated affects the properties of the polymer that is produced. This is about 11% in cellulose nitrate for plastics, and 12% for making film. Nitration of about 13% of the –OH groups produces the explosive guncotton.

Both light and moisture cause cellulose nitrate to decay by loss of the nitrate groups. The –OH groups of the original cellulose re-form, and nitrogen oxides are given off. These react with oxygen and moisture in the air to form nitric acid. The acid has two effects. Firstly it acts as a catalyst for further loss of nitrate groups. Secondly it reacts with the main polymer chain, breaking it down into shorter chains and, ultimately, individual sugar molecules. The latter effect is similar to the acid-catalysed hydrolysis of starch, a molecule that is also a polysaccharide.

Interestingly, some cellulose nitrate objects contain zinc oxide used as a filler and as a white pigment. This method was often used to make imitation ivory. The zinc oxide reacts with any acidic degradation products of the polymer and thus prevents the two effects described above. It therefore acts a stabiliser as well as a filler and a pigment.

Q16. Write an equation for the reaction of zinc oxide with nitric acid.

Cellulose acetate (cellulose ethanoate)
This polymer, Figure 11, is related to cellulose nitrate. It is produced from cellulose by reaction with ethanoic anhydride (acetic anhydride). (Ethanoic anhydride is used rather than ethanoic acid because it reacts in the same way as the acid but more vigorously.) Cellulose acetate breaks down in a comparable way to cellulose nitrate, ultimately producing ethanoic acid in moist conditions. Ethanoic acid is a weak acid, but it acts on the polymer in a similar way to nitric acid.

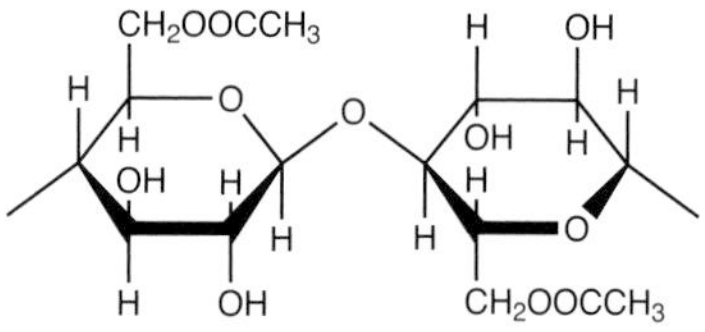

Figure 11 Cellulose acetate

Q17. A conservator might smell a cellulose acetate object that she suspected was beginning to degrade. What sort of smell would she be expecting? Explain your answer.

Q18. Ethanoic acid is a weak acid and will not react as readily with cellulose as will a solution of the same concentration of nitric acid, which is a strong acid.

a) Explain carefully the difference between strong and weak acids.

b) Ethanoic anhydride is used as an alternative reagent to ethanoic acid for making cellulose acetate. It reacts more readily with the –OH groups of cellulose than ethanoic acid but produces the same product. Suggest another alternative reagent with similar reactivity to ethanoic anhydride.

Other reactions of decay products

Except as an academic exercise, the decay of plastics cannot be considered in isolation. An object may not be made of just one plastic – it may contain two or more types of polymer and/or it may contain other materials such as metals. A doll is a good example. The body may be made of one polymer and the hair of another. It may have jointed limbs with metal hinges and it may be dressed in a variety of other materials. It may be stored in packaging made of yet another type of plastic. All these materials may interact with one another.

Did you know? Collectors of many objects that were sold in packaging usually like to collect the packaging too. Barbie™ dolls are now becoming collectors' items. They fetch much higher prices if they are still in the original packaging.

So, for example, acid decay products from a plastic may react with the metal parts of the joint, or vapours given off by one plastic may react with another plastic in the same museum display case. This clearly has implications for the storage and display of plastic objects and means that museum conservators need to be able to identify different types of plastic and to understand the chemistry of the decay processes that take place in these plastics. Without this knowledge, it would be easy to do more harm than good. 'Commonsense' might suggest that storing a plastic object in a sealed container might keep it from harm. However, if the object is made from, say, pvc, this would keep hydrogen chloride gas given off by the object in contact with it and thus catalyse further decay.

RS•C

Conservation of plastics

Modern synthetic polymers have a history going back not much more than 150 years, although the use of natural plastic materials goes back much further, see *A brief history of plastics*.

We do not often think of objects made of plastic as being collectors' items. This is probably for two reasons. Firstly plastic materials tend to be mass-produced and therefore lack rarity value. Secondly, plastics are often regarded as cheap materials and therefore not as worthwhile to collect as more expensive items.

However, there is now an increasing interest in collecting objects made of plastic. This may be from individuals who collect as a hobby, or institutions such as museums that are concerned with preserving and displaying objects because of their historical value, as examples of design *etc*.

The types of objects that might be collected and the reasons for collecting them vary enormously. The spacesuits used on the first Moon landing are of obvious importance historically, and represent the cutting edge of the technology of the late 1960s. However, something as apparently mundane as a carrier bag might represent the design and marketing strategies, and even art, of its period. Some people collect things just for the joy of collecting. For example, Barbie™ dolls have recently become highly collectable and can change hands for as much as £8000.

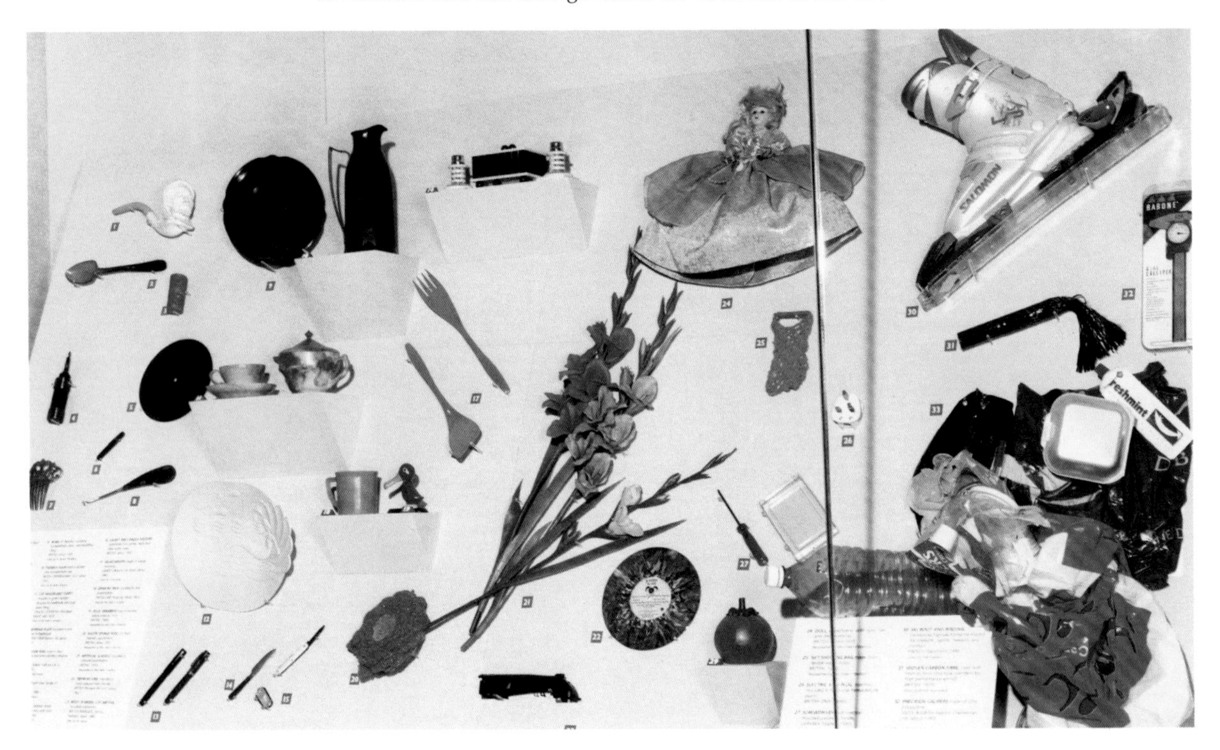

Figure 1 Display case at the National Museum of Scotland
(Picture reproduced by courtesy of the National Museum of Scotland.)

Collectors are obviously concerned to keep the items in their collections in good condition, not least because they have paid good money for them. Museums have the added responsibility to display objects to the public and to preserve them for research and study; the need to display may cause problems with regard to conservation.

Q1. Suggest some ways in which the need to display a plastic object to the public might clash with the need to conserve it.

Despite the popular misconception that they last forever, plastics do decay and deteriorate. One complicating factor is that plastic is not a single material - there are about fifty basic types whose properties can be tailored in various ways to give many thousands of materials with different properties. Many of these tailoring methods involve adding other substances to the base polymer from which the plastic is made. This leads to many more possible chemical reactions by which the material might decay. The situation is further complicated by the fact that many objects will be made of more than one type of material (both plastic and non-plastic).

Q2. The main body of a Barbie™ doll is made from pvc.
Suggest what other materials might be used in making the rest of the doll and its clothing. Try to suggest what material might be used for what part.

Figure 2 Barbie™ dolls

Identifying plastics

Some of the ways in which plastics decay are discussed in *The decay and degradation of plastics.* Before knowing how to care for a plastic object, it is essential to know as much as possible about what type of plastic it is made from and what additives (fillers, pigments, stabilisers *etc*) it contains. How this is done will vary depending on the circumstances and the type of object. For example, an individual collector will not have access to the same facilities for testing as will a national museum. Another issue is that some methods of testing require a sample of the material to be taken from the object (destructive testing) and some do not (non-destructive testing). The size of the required sample will vary from method to method. How acceptable this sampling is may depend on the size of sample required in relation to the size of the object.

Identifying a plastic is very much a case of detective work, with many techniques being used to give clues. Except for some instrumental techniques, it is rare that a single method will give a definitive answer. Often a particular method will rule out certain plastics rather than positively identify the actual material.

Outward appearance

Important clues can come from simply feeling or looking at the object, often with the aid of a magnifying glass. Many objects will have markings such as trademarks, patent numbers and registration marks, Figure 3. These may identify the manufacturer and give an idea of the date of the object. Collectors may then be able to check records and find out that a particular manufacturer made this type of object at that date out of a certain type of plastic. The date of manufacture can also be used to rule out certain plastics that had not then been developed or come into regular usage. For example Nylon did not come into use until about 1940, so objects reliably dated as being significantly older than this cannot be made of Nylon. Even an object without markings can be dated by its style and design – the Art Deco style, Figure 4, came in in the 1920s, for example. However, some care is needed – an Art Deco object could not have been made *before* the 1920s but it could be a newer imitation.

Figure 3 A manufacturer's mark on a plastic item

Figure 4 Objects in the Art Deco style

Colour, and transparency, can help to identify plastic types. Relatively few types of plastic can be made into wholly transparent mouldings and a number of types can only be opaque. Table 1 gives some general guidelines although they need to be used with care – the thickness of a sheet or moulding can affect transparency, for example as can colourants and fillers.

RS•C

Transparent when moulded*	May be either transparent or opaque	Always opaque
Perspex™	Cellulose nitrate	Phenol-formaldehyde
Cast phenolic	Cellulose acetate	Urea-formaldehyde
Polyester	pvc	Vulcanised rubber
Polystyrene	Polypropylene	Bois durci
Polycarbonate	Polythene	Vulcanite
		Shellac

* Addition of pigments can make these appear opaque.

Table 1 Transparency and opacity of some plastics

Colour can also be helpful, if only in ruling out possibilities – phenol-formaldehyde is almost always in dark shades, for example. Surface finish may also give an indication of plastic type – for example, acrylics, polystyrene and cellulose esters can be highly polished, whereas the surface of polythene is much less glossy.

Finally the method of manufacture may help. It is often possible to determine the method by which a plastic object has been moulded from tell-tale markings. Certain plastics are made by only a limited range of techniques. So this can give clues as to what material an object might or might not be made from. For example, injection moulding involves forcing molten plastic under pressure into a mould. This leaves behind a small 'tail' of plastic, called a sprue, where the filling hole was. When this is removed, it leaves a small surface imperfection (see Figure 5).

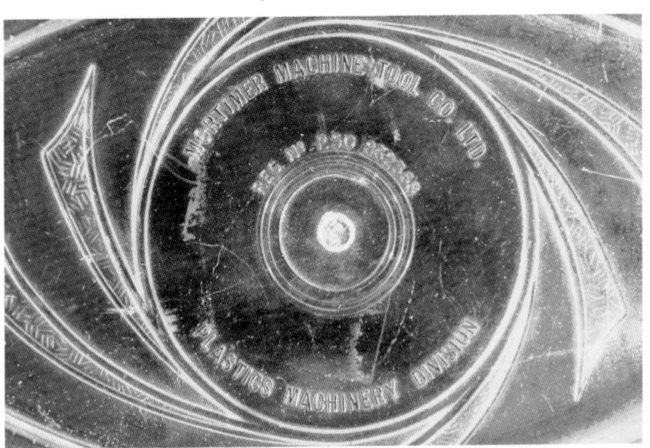

Figure 5 The injection moulding mark on a plastic object

Objects made from casein and from cellulose nitrate were never made by injection moulding so the appearance of this type of mark in a moulding would rule out these types of plastic.

Physical tests

Physical tests are ones in which no chemical change is involved. Two useful ones for helping to identify plastics are measurements of hardness and of density. Perhaps the simplest hardness test is to see if the plastic can be marked by thumbnail. Plastics that can be marked include: polythene, polypropylene, polyurethane, pvc (with plasticiser) rubber and gutta percha. This is clearly a 'low-tech' test that could be applied by a collector at an antique fair, say.

The density of a material is the mass of 1 cm^3 of it. If the density of a plastic can be measured, it can be compared with the known densities of plastics given in reference books. Care is needed in interpretation as different additives can affect the density considerably.

Q3. Density is defined by the equation density = mass / volume. A plastic chess piece was weighed and found to have a mass of 7.0 g and a volume of 6.5 cm^3.

a) What is its density?

b) Which of the plastic types shown in the Table below might it be made from?

Plastic	**Density / gcm^{-3}**
Expanded polystyrene	0.02–0.06
Polypropylene	0.89–0.91
Low density polythene	0.91–0.93
High density polythene	0.94–0.96
Polystyrene	1.04–1.11
pvc	1.20–1.55
Polyethylene terephthalate	1.38–1.40

c) Why is this method not reliable for making a positive identification of the plastic?

d) Suggest a method of finding the volume of the chess piece.

e) Some plastics are made into foams. An example is expanded polystyrene, used as a packing material. What effect will this have on the density of the plastic?

One of the simplest physical tests that can be carried out on plastics is measurement of melting or softening points. Plastics can be divided into thermosetting and thermoplastic (see *A brief history of plastics*). Thermosetting plastics, such as Bakelite, do not melt at all – when heated they remain solid until they begin to decompose chemically. Thermosoftening plastics, such as polythene, do melt, but rather than melting completely at a particular temperature, they tend to soften gradually over a range of temperatures. This is because a sample of polythene, say, is not a pure substance, but a mixture of many molecules with a range of chain lengths and therefore different melting points. The following plastics soften when placed in boiling water: pvc, polythene, polystyrene, cellulose nitrate, cellulose acetate, vulcanite.

Chemical tests

Traditional chemical tests tend to be destructive and are therefore not used by collectors unless it is possible to take a sample of the object without damaging it. The most used chemical test is to heat a small sample in a small test tube containing a piece of moist indicator paper, measure the pH of any fumes and note their smell. Tables 2 and 3 indicate the results found with some plastics. Some of the smells, of course, are difficult to recognise and describe unless they have been experienced before.

RS•C

Safety note

Burning plastics and smelling any fumes produced can be dangerous. It should not be tried without the close supervision of a chemistry teacher.

Plastic	Smell
Casein	Burnt milk or hair
Cast phenolic	Phenol or carbolic soap
Cellulose acetate	Vinegar, burning paper
Cellulose nitrate	Camphor, then nitrogen oxides
Gutta percha	Burning rubber
Melamine formaldehyde	Fishy
Nylon™	Burnt hair, celery
Pet (Polyetheneterephthalate)	Burnt raspberry jam, sweet
Phenol-formaldehyde	Phenol or carbolic soap
Poly(methylmethacrylate) (Perspex™)	Sweetish, fruity
Polythene	Wax candles, paraffin
Polypropylene	Wax candles, paraffin
Polystyrene	Marigolds
Polyurethanes	Stinging
Pvc rigid	Hydrochloric acid, chlorine
Pvc plasticised	Hydrochloric acid + aromatic
Urea-formaldehyde	Formaldehyde (methanal), fishy, ammonia
Vulcanite	Burning sulfur/rubber

Table 2 Smells associated with burning plastics

pH 1-4 acidic	pH 5-7 neutral	pH above 8 alkaline
Cellulose acetate	Polythene	Nylons
Cellulose nitrate	Polystyrene	Phenol formaldehyde
Pet	Perspex™	Urea-/thiourea-formaldehyde
Polyurethane	Polycarbonate	Melamine-formaldehyde
Polyester	Silicones	
Pvc	Epoxies	
Vulcanised fibre		

Table 3 pH of vapours given off by heated plastics

Q4. a) Suggest what acidic gases are given off on heating
(i) cellulose nitrate
(ii) cellulose acetate.

b) Suggest why both polythene and polypropylene produce a smell of wax candles and paraffin when heated.

c) The method for this test suggests gentle heating to avoid the samples burning. What gas are all the plastics likely to give off if they burn in a plentiful supply of air? Is this gas acidic or alkaline?

d) What gas are they likely to give off if they burn in a limited supply of air?

Instrumental tests

These have a number of advantages over simple chemical and physical tests. They can often give more definite results that are easier to interpret and they normally require only a small sample – often no more than a few milligrams, the size of a printed full stop. In some cases, no sampling at all is required. However, the instruments themselves are expensive and require an experienced chemist to operate them and interpret the results.

Infrared (IR) spectrometry

This is probably the most useful single instrumental method for identifying plastics. It involves shining a beam of infrared (heat) radiation of a range of frequencies through a sample and onto a detector, Figure 6. Chemical bonds in the sample will absorb radiation at the exact frequencies at which they vibrate. So after the beam has passed through the sample, some frequencies will be missing – which ones will depend on the chemical bonds present in the sample. The results are presented as a graph of the intensity of the radiation detected after passing through the sample against the frequency of the radiation (often presented in wavenumbers in cm^{-1} (1 / the wavelength of the radiation in cm). A dip in the graph (confusingly, this is usually referred to as a peak) shows that radiation has been absorbed at this particular frequency.

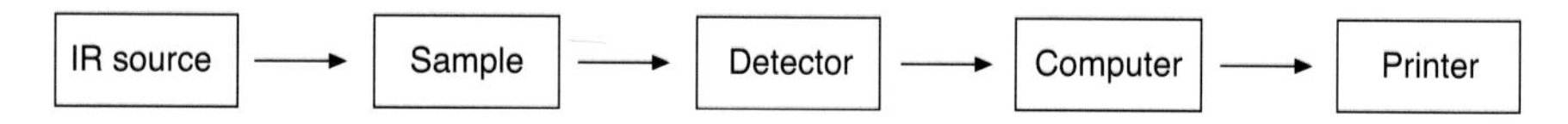

Figure 6 (a) Flow diagram of an infrared spectrometer

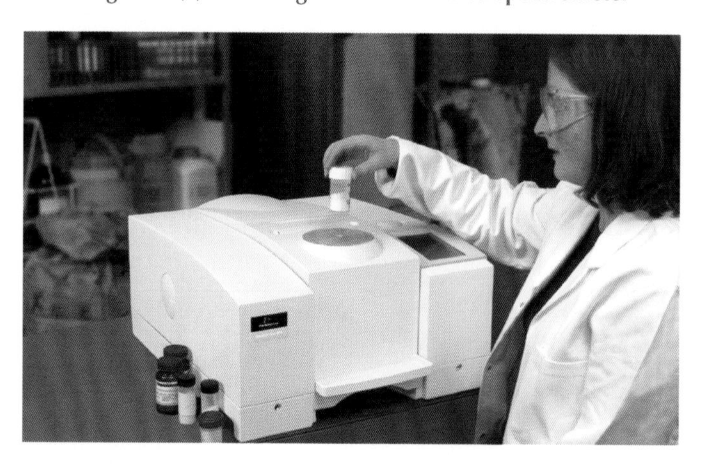

Figure 6 (b) An infrared spectroscopy instrument
(Picture reproduced by courtesy of Perkin Elmer.)

This technique allows the bonds present in the sample to be identified (for example the N–O bond present in, say, cellulose nitrate) absorbs infrared radiation of frequency approximately 1650 cm^{-1}. The infrared spectrum can also be used a 'fingerprint' to identify samples by matching – two samples with identical spectra are almost certain to be the same. Museums analysts are building up collections of IR spectra of known samples of different plastics to be used as references to compare with the spectra of unknown samples. The use of this technique is shown in Figure 7. This shows the IR spectrum of a sample taken from a trinket box next to that of a known sample of cellulose nitrate. The similarity is obvious even without computer matching techniques. The third spectrum is that of a sample of tortoiseshell, another material from which the box might have been made judging by appearance alone. This spectrum is quite different. The absorption at about 1650 cm^{-1} represents the vibration of the N–O bond in the nitro group.

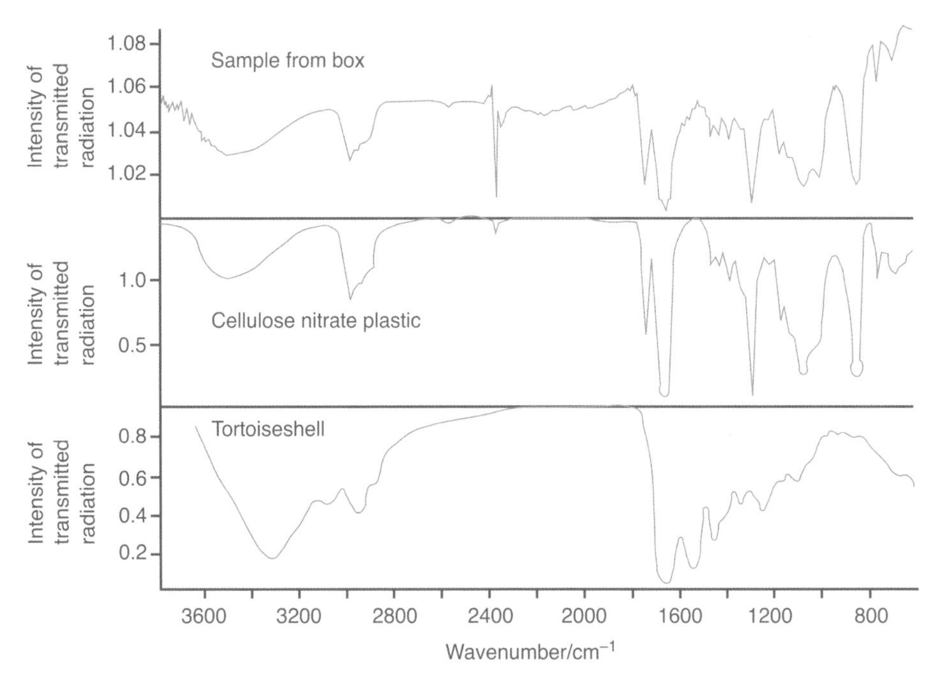

Figure 7(a) Identifying plastic by matching the infrared spectrum

Figure 7(b) The box from which the samples were taken

Gas Chromatography – Mass Spectrometry (GCMS)

This method is really two techniques in one. The gas chromatograph separates the different components in the plastic sample and the mass spectrometer helps to identify them. A solution of the sample in an organic solvent is made and it is injected into a column, a long, thin spiral tube of silica coated internally with an alkane. A stream of inert gas carries the sample through the column and the more volatile (easily vaporised) components are carried through more quickly than the rest. So the various components of the original plastic come out of the column separately. A chromatogram is produced showing how much of each component is present in the original mixture, see Figure 8.

RS•C

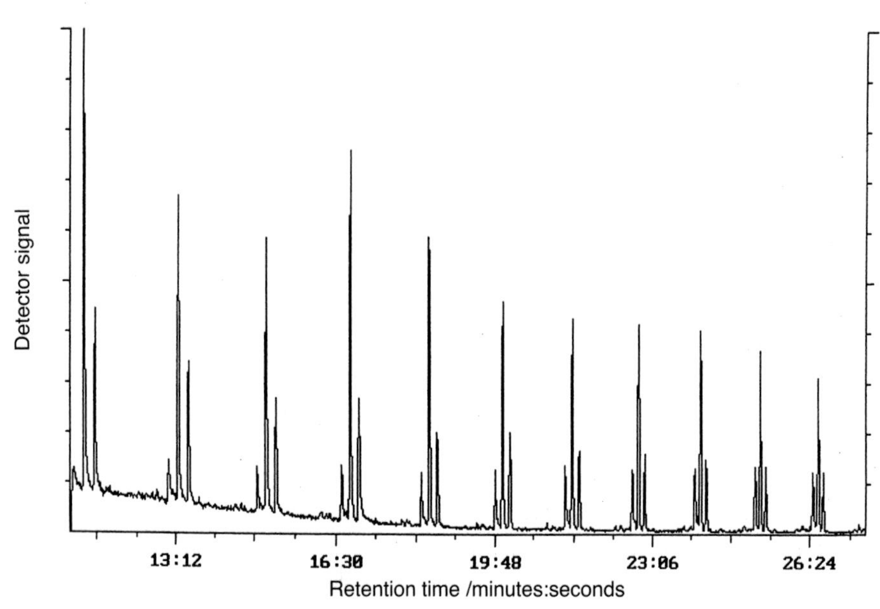

Figure 8 A gas chromatogram of poly(ethene) obtained by heating the plastic. Each trio of peaks represents components of different chain length

(Picture reproduced by courtesy of the Tate Gallery conservation department.)

As it comes off the column, each component is fed directly into a mass spectrometer. Here it is converted into positive ions, most commonly by bombardment in an electron beam. The extra energy given to these ions by the electron bombardment makes them break into fragments. Some of these are charged and fly through the spectrometer, accelerated by electric and magnetic fields. The end result is that the fragments are separated by mass. So each component of the original plastic produces a mass spectrum that can be used as a fingerprint to identify it, see Figure 9.

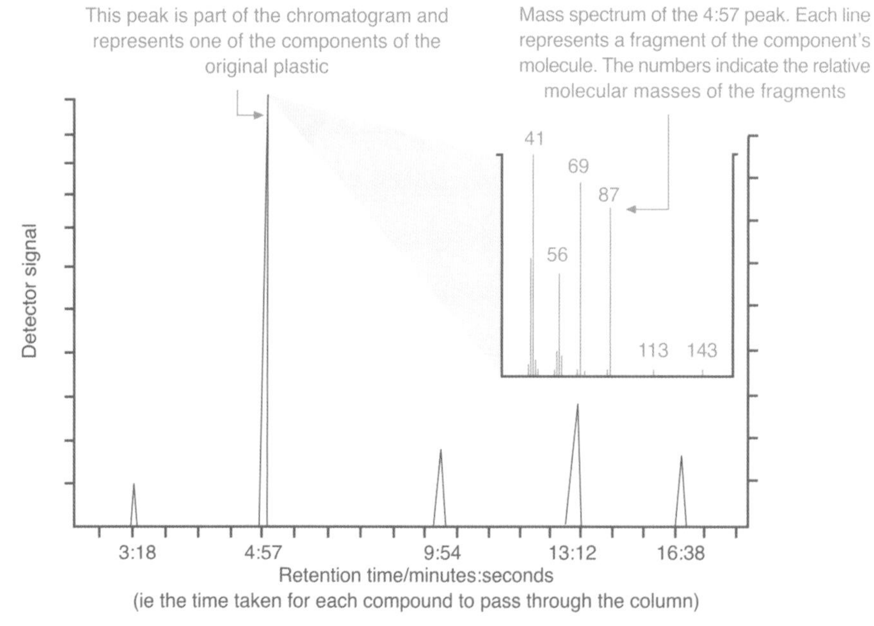

Figure 9 A typical result from GCMS on a sample of plastic

RS•C

Other techniques

Chromatography

Other chromatographic techniques are used – these all work on the same basic principle of separating mixtures on the basis that some components are carried faster through a stationary medium than others.

X-ray diffraction

X-ray diffraction can be used to identify inorganic fillers in plastics. Here a beam of X-rays is fired at a sample and the resulting diffraction pattern is recorded on photographic film, see Figure 10. It is possible to work out from this pattern the actual positions of the atoms in the sample. However, for identifying materials in plastics, the diffraction patterns can simply be compared with those of known samples.

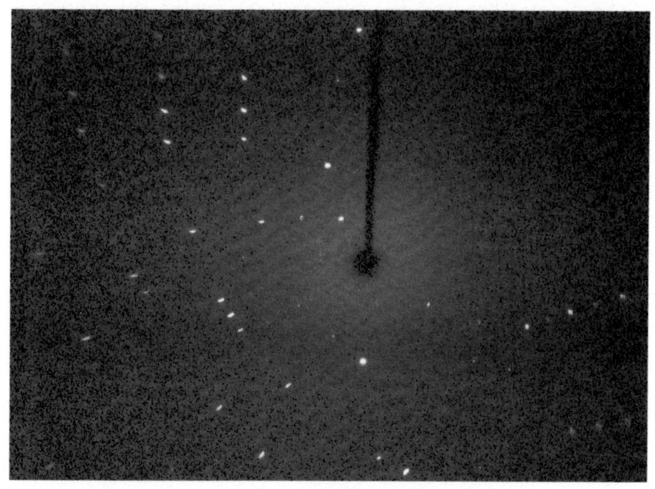

Figure 10 Example of a diffractogram pattern
(Reproduced by permission of the EPSRC National X-ray crystallography service.)

Electron microscopy

Scanning electron microscopy works by firing a beam of electrons at a sample and observing the scattered electrons. It can show the surface at magnifications much higher than optical microscopes. Energy dispersive X-ray spectroscopy is a development of this technique that can identify the elements present in a sample and their amounts. Some of the electrons in the beam have sufficient energy to knock out inner electrons from atoms in the sample. Electrons from outer shells drop into the inner shell to fill the gap and, in doing so, give out X-rays. The frequencies of these X-rays are different for different elements and so the elements in the sample can be identified. If the object to be examined is fairly small, sampling is not needed as the whole object can be fitted into the sample chamber of the instrument.

Q5. **a)** Draw diagrams to show (i) the electron arrangement of a magnesium atom (ii) the electron arrangement when an electron from the innermost shell has been removed (iii) an electron from an outer shell falling into the inner shell to fill the gap.

b) How many possibilities are there for the process of an electron from the outer shell falling into the inner shell to fill the gap to take place? Therefore, how many different frequencies of radiation could be expected?

More detailed information on these techniques can be found in B. Faust, *Modern Chemical Techniques* (1995) and R. Levinson, *More Modern Chemical Techniques* (2001), published by the Royal Society of Chemistry.

RS•C

Case studies

The two cases that follow give some indication about how firstly a private collector, and secondly a museum might go about identifying what a plastic object could be made from.

The individual collector

A collector at an antique fair noticed an interesting amber-coloured decorative box. Simply by looking at its appearance and style and comparing it with similar objects she had seen before, she guessed that it was a trinket box dating from around 1900. This suggested that it was probably made from cellulose nitrate, as this substance was commonly used for making this type of article at that time. The colour supported this conclusion as cellulose nitrate can easily be coloured with pigments.

She picked up the box to look for any manufacturer's markings to confirm her conclusions but there were none. At the same time, she noticed that the box felt heavier than she would have expected if it were made of cellulose nitrate and also that it had moulding lines indicating that it was certainly moulded, and not carved from a natural material.

Slightly puzzled, she rubbed the box vigorously with her handkerchief to warm its surface. Sniffing gently, she noticed a faint odour similar to carbolic soap or TCP antiseptic. This was phenol, which strongly suggested that the box was made from a phenol-formaldehyde plastic such as Bakelite™. This would explain its heaviness. A quick scratch with her thumb left no mark. This is what would be expected for Bakelite™. Perhaps the box was a later copy of a turn-of-the-century piece?

The box was certainly a puzzle but she decided that any further tests to identify the plastic would require a sample to be taken, which the seller was unlikely to agree to, so scratching her head she moved on to the next stall.

A museum

One special dress in the National Museums of Scotland collections was created by couturier Norman Hartnell, who has designed for Her Majesty Queen Elizabeth II. This finely pleated, sequin-covered, silky, low-cut dress was made in 1938-39, but a recent conservation examination found that some sequins were disintegrating.

Box continued...

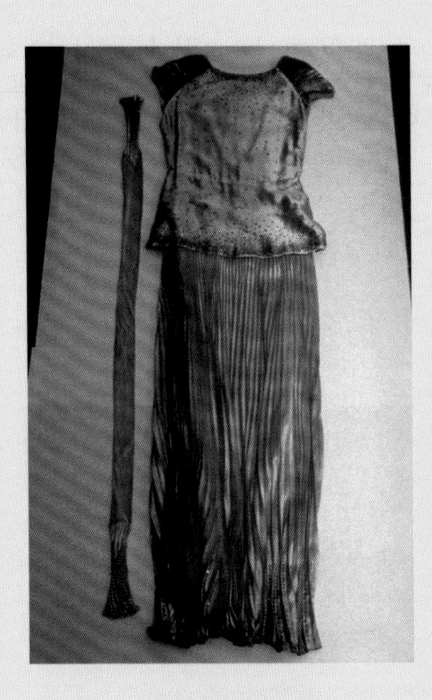

The Norman Hartnell dress with detail showing sequins

RS•C

Samples, the size of printed full-stops, were analysed in the museum by infrared spectrometry to identify what the sequins were made from and therefore possible causes for the deterioration. Analysis revealed that the iridescent bronze-coloured sequins were composed of two man-made plastics: cellulose nitrate applied as a thin layer over casein, a plastic derived from milk protein. The identification was done by running the spectra of the samples and matching them with a computer database of spectra of known plastics.

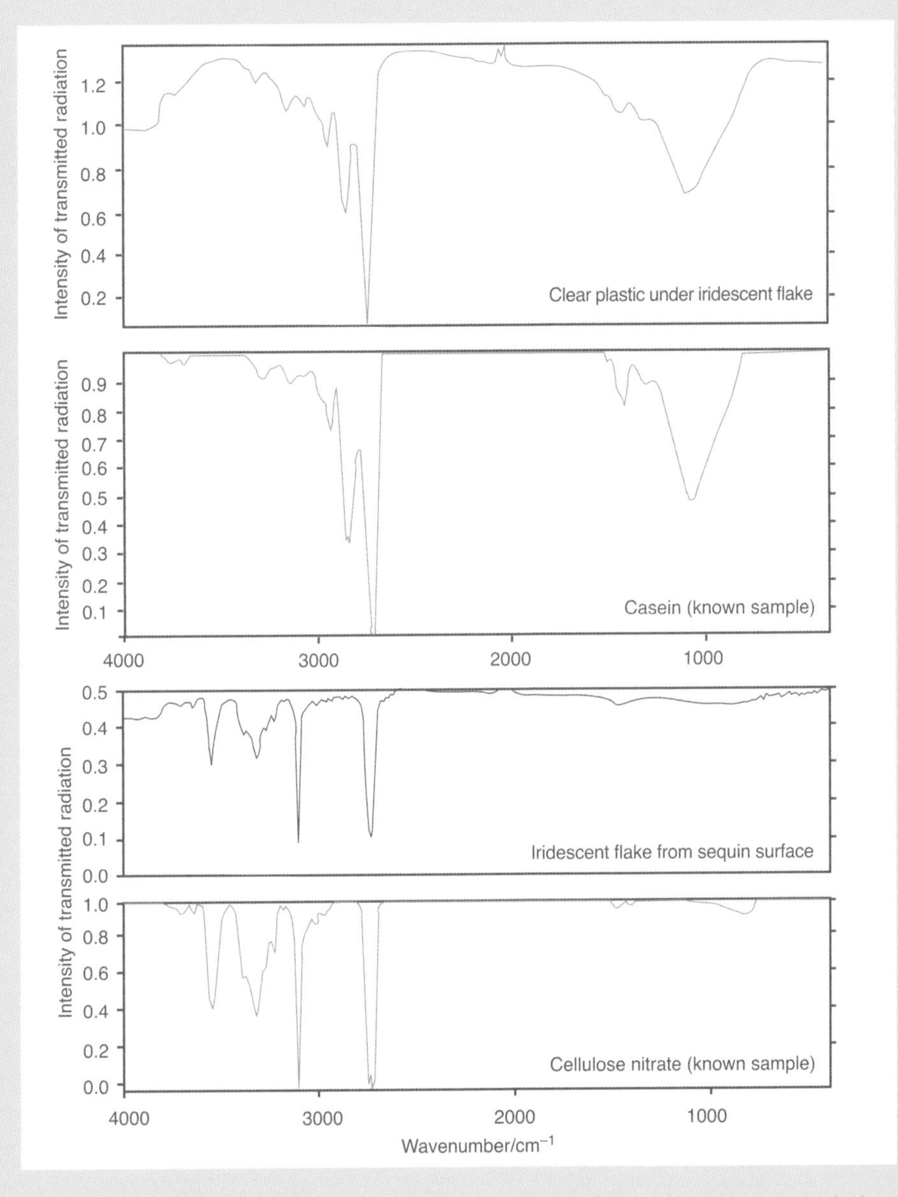

Deterioration could now be explained – these two plastics are incompatible because they behave differently towards moisture. Cellulose nitrate is moisture-sensitive and so it should be stored and displayed in a low-humidity environment. However casein benefits from more humid environments to prevent it from dehydrating and cracking. Inappropriate storage conditions used in the past have caused the two layers to separate as the different plastics contracted or expanded by different amounts. Because the best conditions for each plastic are different, the best compromise has been to maintain stable, average relative humidity (50%) at 20 °C with regular checks on the sequins to see if they continue to deteriorate or, hopefully, stabilise.

Caring for plastics

Here again there may be a difference in requirements between the private collector and a museum. Many private collectors will want objects in their collections to look as new, while museums will often be more concerned to preserve the object from long term damage or deterioration. The difference is often that between so-called passive and active conservation. Passive conservation concentrates of finding conditions for storage and / or display that minimise deterioration while active conservation involves cleaning, repair and restoration.

Passive conservation

We have seen that it is not always easy to identify plastic materials with certainty. This is especially true of modern materials that often contain a complex blend of additives (pigments, fillers, stabilisers, plasticisers *etc*) as well as the basic polymer. Fortunately, many passive conservation measures are applicable to most types of plastic. These include storage at low temperature, away from high levels of light (especially ultraviolet light, present in daylight) and moisture and, for some materials, in the absence of oxygen.

All chemical reactions go more slowly at low temperatures so low temperature storage will slow down all decay reactions. A rough rule of thumb applicable to many reactions is that a temperature drop of 10 °C will halve the reaction rate. Refrigerated storage is not likely to be practical for the private collector and would also create difficulties for items on display. However some archives of cine film, made from cellulose nitrate, which is particularly prone to decompose, are stored at low temperatures.

Q6. How will storing an object at 5 °C (roughly the temperature inside a fridge) rather than at 25 °C (the temperature of a warm room) affect the lifetime of a plastic object?

Q7. Reaction rates are governed by the Arrhenius equation
$k = Ae^{-Ea/RT}$

where k is the rate constant and is proportional to the reaction rate, A is a constant, E_a is the activation energy for the reaction, R is the gas constant, 8.3 JK^{-1} mol^{-1}, and T the temperature in kelvin. The expression $e^{-Ea/RT}$ gives the fraction of molecules that have enough energy to react at that temperature T.

a) The activation energy for a typical reaction is about 50 $kJmol^{-1}$. Using this value of E_a, calculate a value of $e^{-Ea/RT}$ at 25 °C. Remember to convert the temperature to kelvin first and also to convert $kJmol^{-1}$ to $Jmol^{-1}$. You will need to use the e^x button on your calculator.

b) Now repeat the calculation for a temperature of 15 °C (10 °C lower).

c) How do the two fractions compare?

d) Does this explain the rule of thumb given above?

Storage away from light will help to prevent free radical reactions caused by ultraviolet light breaking bonds in the plastic, see *The decay and degradation of plastics.* Glass absorbs ultraviolet light to some extent, so both window glass and the glass of a display case will help to reduce UV. It is also possible to use special types of glass or film coatings that absorb ultraviolet better than normal glass. Objects in storage can easily be kept in the dark, but this is clearly not possible for items on display. However, you may have noticed that many museum galleries do have subdued lighting (as well as being kept cool).

RS•C

A number of plastics decay by oxidation – polyurethane is one example discussed in *The decay and degradation of plastics*. The obvious solution to this problem would seem to be to display items in a sealed environment filled with an unreactive gas such as nitrogen or one of the inert gases. However the need for 100% effective sealing makes this impractical. Moreover, a number of plastics give off acidic gases when they decay and these would tend to build up in a sealed environment. In some cases, such as those of cellulose nitrate and cellulose acetate, these gases can catalyse further reaction and may also affect other materials in the same case. So sealed cases are generally avoided. The effect of storing a cellulose nitrate object in a sealed bag is shown in Figure 11.

Figure 11 The part of this cellulose nitrate fan that has been kept in a sealed bag has clearly deteriorated more than the exposed part

One solution is to use a scavenger, a chemical that reacts with the substance we want to get rid of – in this case oxygen. Some museums use a commercial product called 'Ageless'®, originally developed for prolonging the shelf life of foods, which also decay by oxidation. Ageless® contains finely powdered iron that reacts with oxygen and can reduce the concentration of oxygen to as low as 0.01%.

Q8. **a)** Why is *powdered* iron used in Ageless®?

b) Write a word and a balanced symbol equation for the reaction of iron with oxygen assuming that the product is iron(III) oxide.

c) Use the equation to calculate how much iron would be needed to remove all the oxygen from the air contained in a sealed display case 1 m x 0.5 m x 0.5 m. State what assumption you are making in your calculation.

The case of moisture is not quite so straightforward. In some cases of polymer degradation, water acts as a reactant – the decay of cellulose acetate and cellulose nitrate are examples. So storage in dry conditions makes sense. In other cases, such as casein and some polyesters, which are described as hygroscopic, the plastic itself absorbs water from the atmosphere. If this type of plastic is stored in too dry an atmosphere, it can loose water and begin to crack. The conservator must select and maintain the most suitable water vapour content for the storage atmosphere. The water vapour content of the atmosphere is called the relative humidity – it is the ratio, expressed as a percentage, of the amount of moisture in a sample of air to the maximum amount of moisture that air can hold at that temperature. A relative humidity of between 30% and 50% is recommended for storing most plastics, rising to 60% for hygroscopic ones. Controlling humidity can cause a conflict. We have seen that plastics are best stored in a well-ventilated place to prevent build-up of acidic gases. However, changing the air will make it difficult to keep the humidity constant.

Silica gel is often used to remove excess moisture from the atmosphere. This is a material based on sodium silicate that has many pores throughout its structure in which water molecules can be trapped. When it becomes saturated with water, it can be regenerated by being heated in an oven. A cobalt salt-based indicator is often added, this turns from blue to pink when the gel can absorb no more water.

Stone conservation – statues and antiquities

RS•C

Introduction

This material has been compiled by Ted Lister with the help of the British Museum Department of Conservation.

Acknowledgements

The Royal Society of Chemistry would like to thank the following people for their help in preparing this resource.

Susan Bradley, Head of the Conservation Research Group, British Museum
Vincent Daniels, Principal Scientific Officer of the Conservation Research Group at the British Museum.

The resource

This resource consists of two pieces of material suitable for 14–16 year old students and accompanying Teachers' notes. The first section, *Conserving stone objects – a case study*, discusses some of the issues involved with the museum conservation of objects made of stone and looks at a research project to investigate methods of removing salts from porous stone objects. The second, *Practical work on stone*, follows on from the first and, as well as questions, has some experimental work on identifying types of stone, measuring porosity of stone and investigating fillers or glues. There is a mixture of class practical exercises, planning exercises and suggestions for open-ended investigations.

The *Teachers' notes* contain further details for the teacher, especially about ethical debates about how to conserve objects in museums. It also includes lists of requirements for the practical work and answers to all the questions.

Teachers' notes

How and why we conserve things – ethical debates about how to conserve objects in museums

This section was written by Susan Bradley, Head of the Conservation Research Group at the British Museum. It is presented for the information of teachers who may wish to use it in discussion with their students.

What is conservation?

Museums store and display objects, conserving them to prevent deterioration. To do this work they employ conservators who have been trained in the conservation of collections. A small number of museums employ scientists to carry out scientific investigations on the deterioration and conservation of their collections. These scientists are called conservation scientists, and their work underpins the work of the conservators.

Conservation is about keeping things. In a museum like the British Museum it is about keeping archaeological and historic objects in good condition, and in good repair, so that they can be studied and displayed to provide information and enjoyment for the millions of people who visit the Museum every year, and for future generations. Because archaeological objects, objects that have been dug up from the ground, have often been in wet or salty conditions for a very long time, a lot of damage has occurred. This means that relatively few whole objects are dug up and it also means objects can break down even more when they come into contact with the air. Conservation involves stopping this breakdown and repairing the objects so that they are recognisable.

Conservation is different to restoration. In conservation the conservator reveals the object as it exists, and does not add to it to make it look as it did when it was made. Thus new arms have not been made for the Venus de Milo statue. The conservator is always looking for evidence of how the object was made, or what happened to it when it was in use. This evidence is extremely helpful to the archaeologist, and to those who study the development of ancient technologies.

In restoration an attempt is made to return the object to the way it looked when made. For instance, on a statue lost body parts will be replaced with modern copies. The problem with this is that the copies will be based on studies of other similar statues, and they are not necessarily correct for the individual object. Similarly if a substantial repair is made to a painted and glazed pottery vessel the repair will be painted with a continuation of the design and coated with a resin to give the appearance of the painted, glazed and fired original. This may lead some people to think that the vessel is an intact original, which it is not.

What are the basic principles of conservation?

Conservation scientists, who investigate deterioration and the methods of conservation, and conservators who carry out the conservation of the objects adhere to some agreed principles. These are:

- that the work they do on the objects does not alter the chemistry and physical structure of the material the object is made from;

- that the chemicals, resins and adhesives used in conservation do not increase the rate at which the object decays; and
- that the chemicals, resins and adhesives are themselves stable.

If restoration or repair is to be carried out, it is usual to try to ensure that they are easily removable. So if a chipped porcelain vase is repaired with filler or adhesive, materials should be used that can later be removed without damage to the rest of the piece – ones that are easily soluble, for example. This is sometimes called 'the principle of reversibility'. Here it is vital to understand the chemistry of the materials used – both of the original material and of the filler or adhesive.

How do objects deteriorate?

There are many different materials in museum collections and many different deterioration processes. Many deterioration processes are not fully understood, and new processes are being identified. However there are some ways in which deterioration occurs which are well researched, and materials can be grouped according to these. A few of these are discussed below.

Metal objects corrode in the air through oxidation. The rate of corrosion increases when the air is humid, and when the temperature is high. Metals also react with gases that are present in small quantities such as sulfur dioxide, forming sulfates on the surface. This is the reason that copper alloy statues in the open air are often green.

Some types of stone and pottery objects are porous, that is, water can pass through them. They can often contain soluble salts, such as sodium chloride, which, in the case of pottery, moved into the object during burial, and in the case of stone are either present naturally, or have moved into the stone from ground water. When the humidity in the air changes, the amount of moisture within the pores of the object changes, and the soluble salts start to move. If there is enough moisture present they are in solution in the pores. If the water dries out they crystallise, both within the pores and at the surface of the object, causing damage, particularly at the surface.

Objects made from natural organic polymers such as wood, leather, ivory, bone, paper, cotton, linen and wool react to changes in the moisture in the air, expanding and contracting. If they expand and contract too much, or the changes in dimensions are very large, the objects can be physically damaged with cracks and tears appearing. The polymers also react with water and acids or alkalis in the air, or present within the object, causing hydrolysis (breaking) of the polymer chain. The polymers can also undergo oxidation, again causing breakdown. These reactions cause the polymers to become more brittle and to change colour.

Paints and dyes used to colour objects can fade, or undergo a colour change on exposure to light. Ultraviolet light is particularly damaging, but exposure to visible light alone can also cause fading.

What are the main activities of conservation?

The conservation of an object can involve several different processes:

- examination;
- cleaning;
- repair; and
- stabilisation.

Examination

Objects are carefully examined before conservation begins to ensure that the conservators know exactly what they are dealing with. This may be a visual examination, by eye or under a microscope; or it may involve radiography, analysis on the surface of the object, or analysis of a sample taken from the object. As the conservation process proceeds more examination may be necessary.

Cleaning

Objects may be cleaned because they have become covered with dust over years of storage or display. This type of cleaning is not controversial. Objects may also be cleaned to remove thick layers of corrosion, or deposits that have built up during burial. Some conservators think that such layers should not be removed, or that minimal removal only should be carried out. However for an object to be viewed by the public it must be visible and so this type of cleaning is important in museums. For metal objects cleaning is normally preceded by radiography to identify the shape of the object and provide a guide to the conservator as to what they are likely to find. Removal of corrosion and deposits is called investigative cleaning, and it is normally carried out under a microscope so that any material that may be part of the object, or be evidence of its use, is not discarded with the unwanted layers.

Repair

Objects may need to be repaired for a variety of reasons. They may have been excavated in pieces, or have suffered damage following excavation, or have been previously repaired using an adhesive that has failed. Adhesives of several different types are used to stick together pieces of objects. In selecting an adhesive the conservator will take into account the strength of the material the object is made of, the weight of the object, and the ease of removal of the adhesive.

Another type of repair is consolidation and this is used on materials that are powdering and flaking. This type of deterioration is caused by soluble salts in porous objects, by corrosion of metals, and can be caused by dimensional changes on organic objects. Consolidation involves applying a dilute solution of a resin to the surface of the object to stick the powered or flaking particles to the sound body of the object.

Stabilisation

Many objects in museum collections are stable in the museum environment. For those that are not, stabilisation can be approached in two ways – through a treatment applied to the object, or through treatment of the environment in which the object is stored or displayed (preventive conservation). Treating an object to stop deterioration can involve applying a chemical that reacts with the material the object is made from, chemically altering it. This happens when copper alloy objects are treated with benzotriazole to stop a reaction called bronze disease from occurring. Benzotriazole is the material that is applied to copper roofs to protect them. It makes the roof turn green when it reacts with the copper. On the other hand, silver objects are sometimes coated with a lacquer to reduce the rate at which they tarnish. The lacquer does not react with the silver and can be removed.

Preventive conservation is appropriate when the method by which an object deteriorates is fully understood. It is common to keep metal objects that are prone to corrosion at a low humidity, thus preventing the corrosion reaction from occurring; or to remove pollutant gases, (such as the gases that cause silver to tarnish) from the air by filtration in air conditioning systems. This prevents reaction at the surface of objects. Keeping objects at low temperatures will also slow down deterioration reactions, but this can only be used in storage, not display, because visitors would object to very cold galleries.

RS•C

In order to reduce the deterioration of objects in museum collections conservation scientists conduct research into the materials that objects are made from, and their reactions with the museum environment. It is also important that the materials used in the conservation treatments are fully researched to avoid damaging reactions between the materials the objects are made from and the conservation material. To do their job conservators have to understand the chemistry of the materials they work with.

Conserving stone objects – a case study – Answers

1. Water expands when it freezes; most other liquids contract.
2. Igneous rocks are those formed by the solidification of molten rock. Sedimentary rocks are those laid down by the settling and binding together of particles eroded from other rocks. Metamorphic rocks are those that have been changed by heat and/or pressure.
3. The simplest suggestion is to find the mass of a sample of dry rock, immerse it in water and periodically re-weigh it until there is no further increase in weight (taking suitable precautions to remove surface water).
4. After each poultice is removed, squeeze out some of the water and test for ions present using standard tests such as the formation of a precipitate with silver nitrate to show the presence of chloride ions. Continue until the tests show no ions in the water samples. One source of information about these tests is K. Hutchings, *Classic Chemistry Experiments*, London: Royal Society of Chemistry, 2000, pp 203–207.
5. To ensure that all the soluble salts had dissolved completely.
6. The powder would have a greater total surface area than a lump – this would speed up dissolution of the salts.
7. We need to know the amount of salts in a known mass of the dust in order to make fair comparisons between the samples.
8. EA646. This suggests that the previous treatment of EA1332 had been at least partially successful in removing salts.
9. Calcium ions could have come from the limestone itself (mainly calcium carbonate). Ammonium ions could have come from fertilisers (such as ammonium nitrate) applied to the soil. Other suggestions are possible.
10. Ethanoic acid.
11. See Figure 1 overleaf.

RS•C

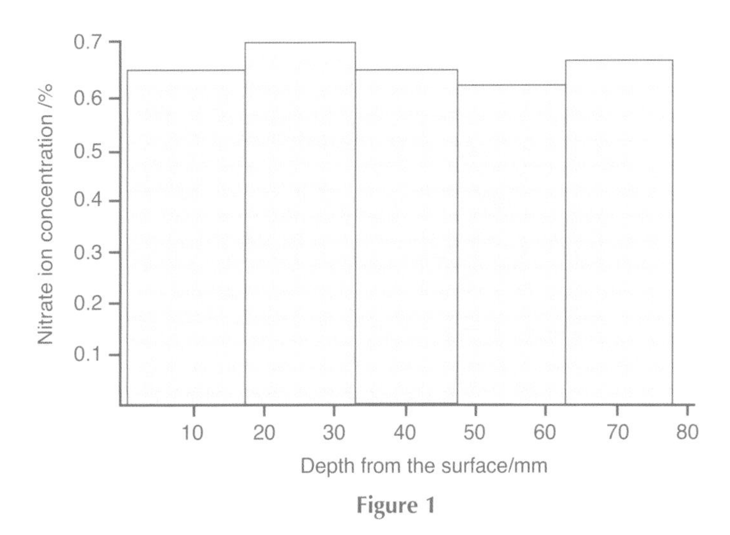

Figure 1

The concentration of ions is almost constant throughout the stela. In particular, there is not a high concentration near the surface and lower ones inside the object.

12. See Figure 2

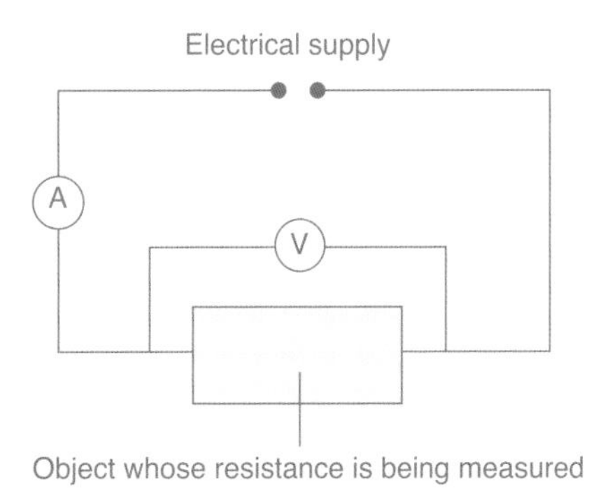

Figure 2

Strictly, the electrical supply should be alternating current to prevent electrolysis of the solution but this is probably too subtle a point for most students at this level. Note that the ammeter must be in series and the voltmeter in parallel.

13. Experiments to measure the water content directly and the resistance of stone samples to produce a calibration curve.

Practical work on stone – notes on the specific experiments

It is the responsibility of teachers to carry out an appropriate risk assessment for all practical work.

Experiment 1 Testing for a carbonate

Apparatus (per group)

- One test tube
- Test tube rack
- Glass rod

RS•C

Chemicals (per group)
- One small lump of calcium carbonate (a marble chip)
- About 10 cm^3 1 mol dm^{-3} hydrochloric acid (irritant)
- About 10 cm^3 limewater (calcium hydroxide solution)

Experiment 2 Flame testing

Apparatus (per group)
- One cork (any size)
- Bunsen burner
- Piece of nichrome wire about 6 cm long (spares will be needed)
- Mortar and pestle

Chemicals (per group)
- One small lump of calcium carbonate (a marble chip)
- About 10 cm^3 1 mol dm^{-3} hydrochloric acid (irritant)

Another type of mineral that could be tested is malachite. This contains copper carbonate and therefore will fizz with acid and give a blue-green flame colour. Many types of sandstone will also contain calcium carbonate.

Experiment 3 Insoluble material

This activity could be used simply as a planning exercise. If it is to be carried out, a sample of sandstone that is soft enough to be easily crushed will be required. The sandstone would need to be crushed in a mortar and pestle and a weighed sample left in hydrochloric acid, possibly with gentle heating, for some time to dissolve any carbonates. The residue is filtered off, washed, dried and reweighed. Strictly, the process should be repeated until there is no further weight change in the residue but lack of bubbles could be taken as a simpler indication that the reaction is complete.

Since the available samples will vary from area to area, teachers are advised to try the experiment beforehand to determine the sort of results to be expected.

Exact requirements for apparatus and chemicals will depend on the plans proposed by the students.

Experiment 4 The reactions of marble with acids

Apparatus (per group)
- One 100 cm^3 conical flask with a bung and flexible delivery tube
- One 100 cm^3 beaker

Chemicals (per group)
- About 20 cm^3 1 mol dm^{-3} hydrochloric acid (irritant)
- About 20 cm^3 1 mol dm^{-3} sulfuric acid (irritant)
- About 20 cm^3 1 mol dm^{-3} nitric acid (irritant)
- Small lumps of calcium carbonate (marble chips)

Experiment 5 The porosity of stone
This activity could be used simply as a planning exercise. If it is to be carried out, small samples of different types of stone will be needed along with access to a top pan balance (reading to at least 0.01 g) and an oven set at about 100 °C.

The principle of the ink soaking into stone can easily be demonstrated by using a stick of suitable blackboard chalk instead of the stone. Different brands of chalk vary; in general, cheaper brands seem to work better than the chalk provided in most schools.

Exact requirements for apparatus and chemicals will depend on the plans proposed by the students.

Since the available samples will vary from area to area, teachers are advised to try the experiment beforehand to determine the sort of results to be expected.

Experiment 6 Investigating fillers
This activity could be used simply as a planning exercise. If it is to be carried out, a variety of different brands of filler will be required (from a DIY store) along with containers and stirrers for mixing (disposable containers and stirrers such as old yoghurt pots and lollipop sticks are a good idea). Samples of types of stone will be needed to test adhesion.

Exact requirements for apparatus and chemicals will depend on the plans proposed by the students.

Since the available samples will vary from area to area, teachers are advised to try the experiment beforehand to determine the sort of results to be expected.

Experiment 7 Investigating glues
This activity could be used simply as a planning exercise. If it is to be carried out, a variety of different brands of adhesive will be required (from a DIY store). Samples of types of stone will be needed to test adhesion.

Exact requirements for apparatus and chemicals will depend on the plans proposed by the students.

Since different glues will vary considerably in their properties, teachers are advised to try the experiment beforehand to determine the sort of results to be expected.

Care should be taken with solvent-based adhesives as the solvent vapours may be flammable and, in any case, should not be inhaled.

Answers to questions

1. Some possibilities include: dropping it; exposure to light; exposure to moisture in the atmosphere; exposure to acidic gases in the atmosphere; exposure to gases given off by other exhibits (especially if both are in the same display case); people touching it; expansion and contraction caused by changes in temperature.
2. (i)
   calcium carbonate + hydrochloric acid → calcium chloride + carbon dioxide + water

   (ii) $CaCO_3(s) + 2HCl(aq) \rightarrow CaCl_2(aq) + CO_2(g) + H_2O(l)$
3. The limewater goes milky. (This is due to the formation of a precipitate of insoluble calcium carbonate, addition of further carbon dioxide will dissolve the precipitate as soluble calcium hydrogencarbonate.)
4. (a) Sodium chloride (common salt) among others.
   (b) This will colour the flame orange as a result of the presence of sodium ions.

RS•C

5. (a) It checks that there is no contamination of the hydrochloric acid solution with ions that might colour the flame.
   (b) If a flame colour does appear, discard the acid and use a fresh supply.

6. They should do.

7. (a) Approximately 10 cm^3.

   (b) 10/24 000 = 1/2400 mol

   (c) 1/2400 mol

   (d) 100 g

   (e) 100 /2400 = 0.04 g

   (f) Probably.

   Note: the other answers will, of course, vary depending on the estimate of the volume of a test tube.

8. See Figure 3.

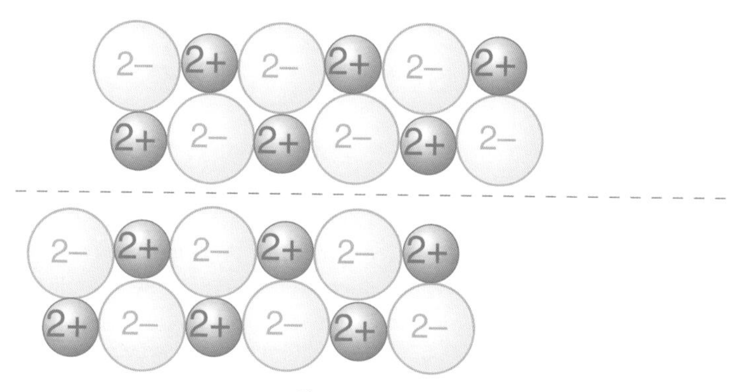

Figure 3

   (b) Ions of the same charge.

   (c) The structure will break apart due to the repulsion of ions of like charge that are touching.

9. $S(s) + O_2(g) \rightarrow SO_2(g)$

   $2SO_2(g) + O_2(g) \rightarrow 2SO_3(g)$

   $SO_3(g) + H_2O(l) \rightarrow H_2SO_4(l)$

10. (a) Sulfuric acid

   (b) (i) calcium nitrate (ii) calcium sulfate (iii) calcium chloride

   (c) (i) calcium nitrate: approximately 1000 g dm^{-3} (ii) calcium sulfate: approximately 6 g dm^{-3} (iii) calcium chloride: approximately 600 g dm^{-3}. Calcium sulfate is the least soluble. These figures will vary a little according to source but their relative magnitudes should be clear.

   (d) The reaction will slow down and eventually stop.

11. Water expands when it freezes. If the water were to freeze, this expansion could cause the statue to crack. If a bottle of milk freezes, the water expands and will push off the top.

RS•C

12. Suggestions might include: adhesion to statue; non-porous; similar colour to original material; able to be removed easily (with a suitable solvent, for example); expansion and contraction similar to the original material.

13. Suggestions might include: adhesion to statue; non-porous; similar colour to original material (or colourless); able to be removed easily (with a suitable solvent, for example); expansion and contraction similar to the original material.

Further notes

Marble, limestone and calcium carbonate

These three terms are used throughout the resource and it may be worth ensuring the students understand the differences between them. Both limestone and marble are largely calcium carbonate. Marble is a form of limestone that has been metamorphosed – that is, its structure has been changed by heat and pressure. It is harder than limestone.

The brittleness of stone

The explanation of the brittleness of stone given in **Practical work on stone** is probably an oversimplification for many types of stone. The explanation in terms of layers of ions sliding over one another is probably accurate for pure ionic materials such as marble but in most types of stone, the situation is more complex. Typically the stone may consist of small crystals of one material cemented together by other materials and breakage will probably occur between these crystals, Figure 4.

Figure 4 Limestone – 'large' calcite crystals in a 'cement' of calcite and clay material

Sulfur oxides in the atmosphere

Care should be taken not to give the impression that all the oxides of sulfur in the atmosphere result from the actions of mankind. Volcanic gases are a significant source of sulfur dioxide in the atmosphere – this gas typically forms 8% of volcanic gases.

Many fuels, such as petrol and diesel, are treated during the refining process to remove sulfur compounds and many power stations have a flue gas desulfurisation plant fitted.

Adhesives

Adhesives work in a number different ways. The essential principle of many of them is that the two materials being joined are held together by intermolecular forces. The relative weakness of these forces is compensated for by the large number of them that operate when significant areas of the two surfaces approach very closely. Adhesives wet both surfaces and then set hard by either solvent evaporation or a polymerisation reaction. This leaves both surfaces in very close contact with the now solid adhesive allowing the intermolecular forces to hold them together.

Some other types of adhesive form actual covalent bonds between the two surfaces being joined.

Conserving stone objects – a case study

We all know that museums store and display historical objects. This is so that the public can see them and also so that historians and others can use them for research. Museums need to preserve these objects in the best possible condition to prevent them deteriorating. This is the job of the conservation scientist, who needs to understand how different materials decay, and devise methods to slow down deterioration.

Many objects in museums are made of stone, Figure 1. Most types of stone seem to be strong and chemically fairly unreactive – you would probably not think that a stone object would need action to conserve it. However, a stone object can be damaged in a number of ways, Figure 2. These include, for example:

- it could be dropped which might chip or shatter it;
- it could be attacked by acidic gases; or
- it could soak up water which could freeze and shatter it.

Figure 1 Statue of the Three Graces by Canova on display at the Victoria and Albert Museum

(Picture reproduced with permission from the V&A Picture Library.)

Figure 2 The Aberdeen Head, a Greek statue dating from about 325–280 BC

(Reproduced by courtesy of the British Museum.)

Q1. What unusual property of water is illustrated by the fact that when water freezes it can shatter stone?

Salts in stone

Water soaking into stone can damage it in another, less obvious, way. Water can carry soluble salts into the pores of the stone. These salts can crystallise in the pores and, as the crystals form, they may exert pressure and crack the stone. If a stone object such as a statue has been buried for some time, salts are often carried into the stone from the soil. This case study looks at the problem of salts in stone and how to remove them. It describes a research project into how to remove these salts that was carried out by conservation scientists at the British Museum in London.

Types of stone

Stone can be classified into one of three types – igneous, metamorphic and sedimentary. These differ in how porous they are (that is how much water they can soak up). Igneous rock is generally non-porous and does not present problems due to water soaking into it. Most types of metamorphic rock are slightly porous, typically absorbing about 5% of their own volume of water. Sedimentary rocks are generally more porous and may absorb between 15 and 30% of their own volume of water.

Q 2. Explain briefly the terms igneous, metamorphic and sedimentary.

Q 3. Devise an experiment to measure the porosity of a sample of rock. Explain what you would do and what measurements you would make.

What is a salt?

Salts are compounds in which a metal has replaced the hydrogen of an acid. For example zinc sulfate ($ZnSO_4$) contains the metal zinc that has replaced the hydrogen in sulfuric acid (H_2SO_4). The metal is obvious from the name of the salt, and the acid usually is too – nitrates are related to nitric acid, sulfates to sulfuric acid and so on. Common exceptions, however, are chloride salts such as sodium chloride (NaCl). Their parent acid is hydrochloric acid. The difference in the names is because the ending '–ate' means 'containing oxygen' while '-ide' means 'and nothing else'. Sulfate and nitrate ions contain oxygen but chloride ions do not.

Salts are all ionic compounds. The metal forms a positive ion, and the negative ion comes from the acid. Some common negative ions found in salts are:

Sulfate, SO_4^{2-};

Nitrate, NO_3^-; and

Chloride, Cl^-.

The only common positive ion found in salts that is not derived from a metal is the ammonium ion, NH_4^+.

Removing salts from stone objects

Crystallised salts can often be seen on the outside of stone objects – you can sometimes see them on bricks, for example, see Figure 3.

RS•C

Figure 3 Efflorescence (salt deposits) showing up as white patches on a house and a wall
(Pictures reproduced by courtesy of the Brick Development Association www.brick.org.uk)

Salts on the outside can damage the surfaces of objects, but they are not the major problem for conservation scientists. This is because salts on the outside can usually be removed by brushing. It is also easy to see when they have been removed. However, in order to prevent damage to a stone object, conservation scientists may need to remove any salts that have been absorbed inside it. This can be more difficult. One method of doing this is to surround the object with filter paper soaked in pure water. This is called poulticing, and the water-soaked filter paper is called a poultice. Water from the poultice soaks into the object and dissolves any soluble salts. The poultice is then dried, and this draws the water (now containing dissolved salts) back into the poultice. This is removed (along with the salts it contains) and replaced with a fresh poultice soaked in pure water. The process is repeated many times.

Q4. One problem with poulticing is how to know when all the salt has been removed. Suggest how this could be done. Hint. There are a number of simple tests for different salts such as chlorides, sulfates, nitrates *etc*. You could look these up in a chemistry textbook or database.

Researching the process

Conservation scientists at the British Museum wanted to investigate the poulticing process to see how effective it was at removing salts and to find out whether previous treatment of the stone affected the efficiency of the removal process. They also wanted to know how far salts penetrated into stone objects and how far the water from the poultice got. This information would help conservation scientists using this process in the future.

To do this they investigated two Egyptian carvings made from limestone (a type of sedimentary rock). These are called stelae (the singular of this word is stela). They look rather like the ones in Figure 4.

Figure 4 Stela EA1332 (left) and EA646 (right)

These were identified by the museum identification numbers EA1332 and EA646. 'EA' stands for Egyptian Antiquity. EA1332 had been previously treated by conservation scientists because of problems with salts, EA646 had not. Egyptian carvings with high salt contents were already known to deteriorate more rapidly than those containing less salt.

Samples of the stone from the stelae were obtained by drilling into them (from the back so as not to damage the carving). The dust from the drilling was collected at different depths into the stelae – firstly from the surface to a depth of 2 mm and then every 15 mm until the drill bit almost reached the front, carved surface. A weighed sample of the dust at each depth was taken, shaken with 10 cm^3 of pure water and left for 16 hours to dissolve any soluble salts.

Q5. Why was the sample left in water for so long?

Q6. Why was it helpful that the sample was in the form of dust rather than lumps?

Q7. Why was it important to weigh the samples of dust?

The samples were then filtered and the concentrations of various ions in them were measured by a method called ion chromatography. The results for the first 2 mm are shown in Table 1. These results represent the layer of stone closest to the surface.

Negative ions	**Ion concentration/%**		**Positive ions**	**Ion concentration/%**	
	EA1332	**EA646**		**EA1332**	**EA646**
Chloride	1.076	8.588	Sodium	0.673	5.565
Nitrate	0.591	0.528	Potassium	0.098	0.638
Sulfate	0.420	2.721	Ammonium	0.002	0.210
Methanoate	0.046	0.629	Magnesium	0.081	0.125
Ethanoate	0.042	0.392	Calcium	0.644	3.746

Table 1 Concentration of ions found in the two stelae from the surface to a depth of 2 mm/%
(*ie* (mass of ion / mass of stone) x 100)

Most of the above ions may be found in soil, but methanoate and ethanoate are not. The conservation scientists are sure that they have come from the wood of the museum cases in which the stelae had been displayed.

Q8. Which *stela* contained most ions? What does this suggest about the treatment of *stela* EA1332?

Q9. Suggest where (a) the calcium ions and (b) the ammonium ions might have come from.

Q10. What is the name of the acid to which ethanoate ions are related?

Table 2 gives the concentrations of nitrate ions at different depths from the surface of stela EA1332.

Depth from surface/mm	Nitrate ion concentration/%
2–17	0.66
18–32	0.71
33–47	0.66
48–62	0.65
63–77	0.67

Table 2 The concentrations of nitrate ions at different depths from the surface of stela EA1332

Q11. a) Plot a graph of nitrate ion concentration (vertically) against depth from surface (horizontally).

b) What pattern do you see?

The water used in poulticing can only remove salts if it can reach them. So the conservation scientists wanted to know how far the water soaked into limestone. They did this by measuring the electrical resistance of the limestone – the more water the stone contains, the less its resistance.

Q12. Draw an electrical circuit, including an ammeter and a voltmeter, which could be used to measure electrical resistance.

For this experiment they used blocks of limestone rather than the stelae, to avoid damage to the carvings. This was because two holes had to be drilled into the limestone 1 cm apart, each taking a wire placed in contact with the stone at a certain depth from the surface (see Figure 5). An electrical circuit was used to measure the resistance of the stone between the two wire probes.

RS•C

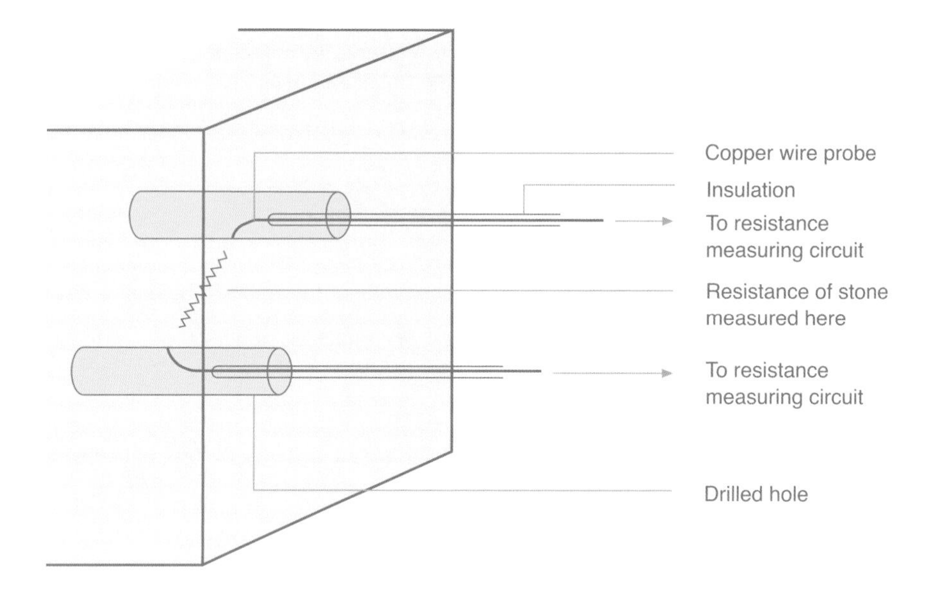

Figure 5 The experimental set-up to measure the resistance of the stone

Three blocks of limestone were tested – one was untreated and the other two had been treated with different silicone-based sealants somewhat like those used to make a waterproof seal round the edges of baths and sinks in the home. A poultice was applied to each block as described above. The resistance of the stone was measured at different depths into the stone over a period of three days. The resistance was then used to calculate the water content of the stone.

Q13. What extra experiments would the conservation scientists have had to do to allow them to convert the resistance figures into percentage water content of the stone?

The results indicated that water penetrated up to 40 mm into the untreated stone but less than 20 mm into the treated stone.

Three significant pieces of information resulted from these experiments:

1. Water from the poultice does not penetrate all the way into stone objects, so there is a limitation on how useful the poultice method of salt removal can be.
2. There is a fairly even distribution of salts throughout the stone – salts are not just confined to a layer close to the surface.
3. The previous treatment had reduced water penetration into the stone.

In the future, conservation scientists will be able to use these conclusions to help them plan the treatment of other stone objects.

RS•C

Practical work on stone

Museums store and display historical objects. Part of the job of a conservations scientist in a museum is to slow down the deterioration of such objects. Imagine you have a statue made of marble that is going to be displayed to the public in a museum.

Q1. Make a list of all the possible ways in which the statue could be damaged or begin to deteriorate if it were exhibited in a museum. Hint. What sort of things is the statue exposed to? What could happen to it? What sort of changes will occur around it during the day and night?

Before we can think about how to conserve the statue, we will need to know what type of stone it is made from. One of the most common types of stone for making statues is marble. Others include sandstone and granite, see Figure 1. We can use the appearance of the stone and chemical tests to help us decide if the statue could be made of marble.

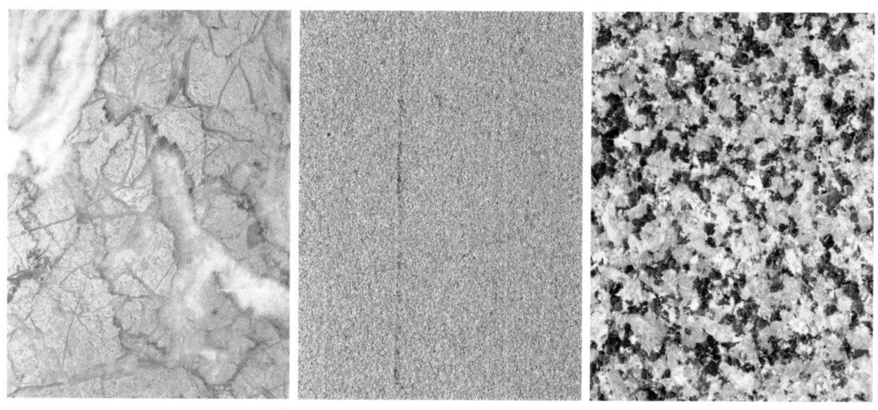

Figure 1 left to right: marble, sandstone & granite
(Reproduced with permission from Science Photo Library. Photographer George Bernard.)

Testing for marble

Marble is hard and is usually white, but impurities can give it a variety of colours, including black. Marble is made of calcium carbonate, $CaCO_3$. All carbonates react with acid to give off the gas carbon dioxide so this can be used as a test for calcium carbonate.

Q2. Write (i) a word and (ii) a balanced symbol equation for the reaction of calcium carbonate with hydrochloric acid.

Q3. We can test for carbon dioxide by using limewater (calcium hydroxide solution). Describe the result of this test.

Experiment 1 Testing for a carbonate

You can try the test for a carbonate on a small chip of marble. Place a marble chip in a test tube and add about 1/3 test tube of 1 mol dm^{-3} hydrochloric acid. You should see bubbles of gas. Test the gas by holding a drop of limewater on a glass rod in the mouth of the test tube, see Figure 2. After a few moments, look closely at the drop. What do you see?

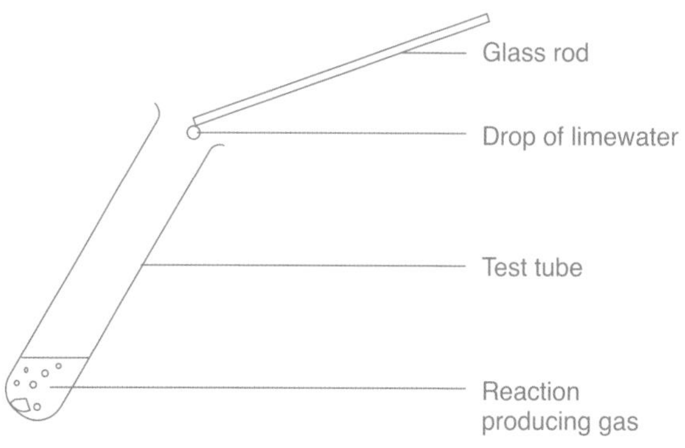

Figure 2 The limewater test

Bear in mind that this test shows only that the stone sample contains a carbonate – marble is almost pure calcium carbonate but some sandstones contain small percentages of calcium carbonate along with other substances such as silicon dioxide.

Experiment 2 Flame testing

We can find out what metal is in a carbonate by using a flame test. When heated in a Bunsen flame, certain metals give particular colours to the flame as shown in Table 1 and Figure 3.

Metal	Flame colour
Calcium	Brick red
Copper	Green-blue
Sodium	Orange-yellow
Potassium	Lilac
Lithium	Scarlet

Table 1 Some examples of flame colours

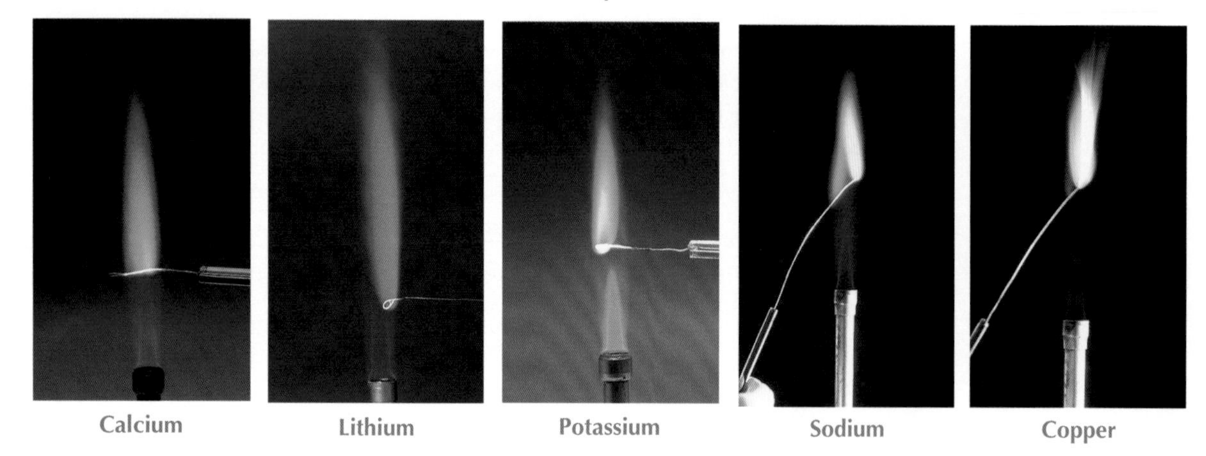

Figure 3 Flame test colours
(Reproduced with permission from Science Photo Library.)

Take a clean piece of nichrome wire about 6 cm long. Push one end into a cork to act as a handle. Hold the wire in the hottest part of a Bunsen flame (just above the blue cone, see Figure 4).

RS•C

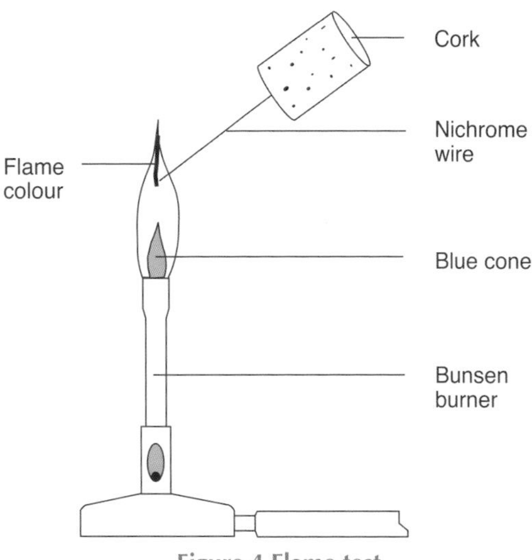

Figure 4 Flame test

If it is clean, the wire should barely colour the flame. If there is a strong colour from the wire, discard it and get a new piece. Avoid handling the wire as it will be hot and when cold it will pick up traces of sweat from your fingers.

Q4. **a)** What salt is present in sweat?

b) What colour will this turn the flame?

Now dip the wire into a little 1 mol dm^{-3} hydrochloric acid and again place it in the flame. There should still be no colour in the flame.

Q5. **a)** What is the point of this part of the test?

b) What should you do if the flame becomes coloured when the acid-dipped wire is placed in it?

Grind a small marble chip to powder in a mortar and pestle. Dip one end of the wire into hydrochloric acid and then immediately place the wetted end into the powdered marble. Now place the wire in the flame. What colour does it produce?

Q6. Do the results of your tests indicate that marble contains calcium carbonate?

Q7. (Harder) If a conservation scientist has to test a statue to find out what it is made of, he or she will want to use as little of the stone as possible to avoid damaging the statue. You will probably agree that the limewater test needs more marble than the flame test. This exercise shows you how to estimate how much stone is needed for the limewater test. Note that this is an exercise in estimation. Use round numbers, so that an estimate of the volume of a test tube to the nearest 10 cm^3 would be sensible, for example.

a) Estimate the volume of a test tube. (This is roughly the volume of carbon dioxide that is made for the test.)

b) How many moles of carbon dioxide is this (1 mole of any gas has a volume of about 24 000 cm^3 at room temperature and pressure)?

c) The equation for the reaction of calcium carbonate with acid tells us that one mole of calcium carbonate produces one mole of carbon dioxide. How many moles of calcium carbonate is required for the test?

d) What is the mass of a mole of calcium carbonate, $CaCO_3$, (A_rs: C = 12, O = 16, Ca = 40)?

e) How many grams of calcium carbonate is needed for the test?

RS•C

f) Do you think that this could be taken from a small statue without leaving an obvious mark?

It is worth bearing in mind that more advanced methods of analysis need much smaller amounts of material to test, and that some may be completely non-destructive, that is they can be done on the complete object without having to remove a sample at all.

Experiment 3

Insoluble material

Another test used to find out about stone is to measure the percentage of material that does not dissolve in acid. Plan an experiment to measure this – say what you would do and what measurements you would make.

Damage to stone

One problem with stone objects is that they are brittle – a force such as a small tap can make them shatter. This means that a statue could break into many small pieces or that a part could break off – the Venus de Milo is a famous example, see Figure 5. Either way, the conservation scientist is left with the problem of whether to repair the object or not and, if so, how to do it.

Figure 5 The Venus de Milo statue

(Picture reproduced with permission from Corbis photo library.)

Q8. This question helps you explore why most types of stone are brittle. This is to do with the fact that most types of stone are made of ionic compounds. Take marble (calcium carbonate, $CaCO_3$) as an example; calcium carbonate consists of a giant structure made up of alternating Ca^{2+} and CO_3^{2-} ions, see Figure 6. Notice that positive ions are always surrounded by negative ions and vice versa. The structure is held together by the attraction of these oppositely charged ions with each other.

RS•C

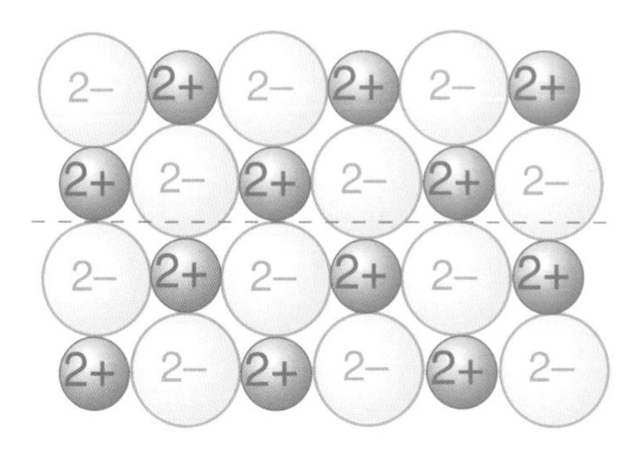

Figure 6 A giant ionic structure

a) A small force can make one layer of ions slide past the next layer. Redraw Figure 6 to show the ions above the dotted line each moving one ion to the right while the ones below the line stay in the same position.

b) What sorts of ions are touching now?

c) What will happen to the structure?

The effect of acids on limestone

You might think that objects made of stone would last forever, but this is far from being true as is shown by the gargoyle in Figure 7.

Figure 7 Badly eroded gargoyle at Whitby Abbey
(Picture reproduced with permission from English Heritage.)

This gargoyle, made from limestone (which is also a form of calcium carbonate), has been eroded by acid rain. Rainwater is naturally acidic because of the reaction of carbon dioxide in the atmosphere with water to form carbonic acid, H_2CO_3.

$$CO_2(g) + H_2O(l) \rightarrow H_2CO_3(aq)$$

So rainwater will dissolve calcium carbonate. As well as erosion of gargoyles, this process is responsible for the formation of caves in areas where the rock is limestone.

Nitric and sulfuric acids are also found in the atmosphere partly because of the burning of fossil fuels. Most fossil fuels contain a little sulfur and, when this is burned, it combines with oxygen to form first sulfur dioxide and then sulfur trioxide. Sulfur

RS•C

trioxide reacts with water in the air to form sulfuric acid. Many power stations now have systems in place to remove sulfur dioxide from their emissions.

Q9. Write word and symbol equations for the reactions by which sulfur in fuel becomes sulfuric acid. You will need the following formulae:

sulfur dioxide; SO_2

sulfur trioxide; SO_3

water; H_2O

sulfuric acid; H_2SO_4

When fuels burn at high temperatures, some of the nitrogen and oxygen in the air combine to produce a mixture of nitrogen oxides. These too combine with water in the atmosphere, to form nitric acid in this case.

Experiment 4 The reactions of marble with acids

Investigate the reactions of marble with nitric acid, sulfuric acid and hydrochloric acid. What gas is produced in each case? Do a test to confirm this. Can you compare by eye the rates of the reactions? Do they all seem to go at the same rate? What factors will you have to keep constant to make this a fair comparison? To confirm your visual comparison of the reaction rates, try the following experiment. Set up the apparatus shown in Figure 8.

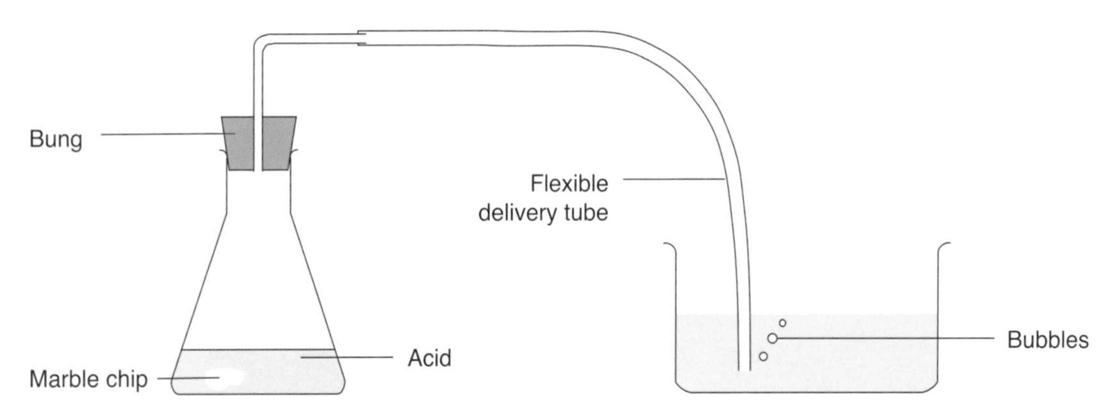

Figure 8 Bubble counting

Add a marble chip and quickly replace the bung and delivery tube. Count the bubbles produced each minute for five minutes. Repeat with a different acid, keeping all other factors as nearly the same as possible. You could plot graphs (possibly using a spreadsheet computer program) of your results. How could you modify the apparatus so that you could measure the volume of gas produced rather than just counting bubbles?

Q10. **a)** You should find that one of the acids reacts more slowly than the others – which one? Try to explain your results by answering the following questions.

b) What salt is produced by the reaction of calcium carbonate with (i) nitric acid, (ii) sulfuric acid and (iii) hydrochloric acid?

c) Look up in a data book or database the solubilities in water of each of these salts – which is the least soluble?

d) What is likely to happen to the reaction of marble with acid if the marble becomes coated with an insoluble salt?

The porosity of stone

Some types of stone are relatively porous, that is they soak up water. This is because they have a network of microscopic cavities, called pores, in their structures.

Q11. How could water soaking into a statue cause it damage? Hint. Think about what happens to a bottle of milk (mostly water) when it is left on the doorstep in freezing weather.

Experiment 5 The porosity of stone

You can compare the porosity of different types of stone in a number of ways. You could immerse a sample of stone in water for different lengths of time and weigh it to find out how much water it has absorbed. You could stand a sample of stone in a beaker of water that has been coloured by adding a little ink and measure how far up the sample the ink has travelled after different time intervals.

Plan an experiment to compare the porosity of different stone samples by one of the above methods. You will need to think carefully about what factors you will keep the same to make the experiment a fair test, what measurements you will take and how to record them. You may need to do preliminary experiments first to test your ideas.

Use of fillers

There is a debate in the world of museums about whether it is right to repair broken objects or not. Some think that this allows the public to see objects as they were originally made, others believe that any damage is part of the history of the object and should be left. Either way, there is agreement that any repair or restoration should be done so that it can be removed at a later date without further damage to the object. This is sometimes called the principle of reversibility.

A simple way to repair chips or cracks in a stone object would be to use a filler like the fillers used to repair cracks in plaster when decorating at home.

Q12. Make a list of the properties that a filler would need to make it suitable for repairing a statue. Think about how this might affect the appearance of the object and about the principle of reversibility.

Experiment 6 Investigating fillers

Plan an investigation to compare the properties of different brands of filler.

Some of the properties you could investigate might include:

- How porous are they? (How much water does a sample of dry filler soak up when it is placed in water?)
- How well do they stick to the material of the statue? (Remember they will need to stick quite well to be any use at all but should not stick so well that they cannot be removed without damage to the object.)
- What is the easiest way to remove them?
- Do they expand or shrink in different conditions such as when wet or when the temperature changes?
- How long do they take to set?

But you may be able to think of other relevant properties.

Use of glues

A simple way to repair a broken stone object would be to use some type of glue to stick back any parts that have broken off.

Q13. Make a list of the properties that a glue would need to make it suitable for repairing a statue. Think about how this might affect the appearance of the object and about the principle of reversibility.

Experiment 7 Investigating glues

Plan an investigation to compare the properties of different brands of glue.

Some of the properties you could investigate might include:

- How strong are they? (What load is required to pull apart two pieces of material that have been glued?)
- How effective are they in different conditions? (Is the strength of the bond affected by temperature or moisture, for example?)
- How well do they stick to the material of the statue? (Remember they will need to stick quite well to be any use at all but should not stick so well that they cannot be removed without damage to the object.)
- What is the easiest way to remove them?
- Is the join visible?
- How long do they take to set?

But you may be able to think of other relevant properties.

Wood conservation – the Mary Rose

RS•C

Teacher's notes

Acknowledgements

The Royal Society of Chemistry thanks Dr Mark Jones and colleagues at The Mary Rose Trust and Dr Des Barker, formerly of Portsmouth University for their help in producing this resource.

The resource

Overview

The Mary Rose is a wooden Tudor warship that sank off Portsmouth in 1545, during the reign of Henry VIII (1509–1547). Most of her hull became covered in silt, which effectively sealed it and the artefacts it contained in anaerobic conditions and preserved them from decay. In 1982, the hull was raised and since then the hull has been undergoing conservation treatment in a former dry dock at Portsmouth. Over 19 000 artefacts were recovered from in and around the ship and these are also being preserved. Many of them are on display in a museum next to the hull. The ship and the objects give historians and the general public a rare insight into life at sea in Tudor times.

The material presented here looks at the chemistry of the decay processes and the methods used to conserve the wood of the Mary Rose's hull and some of the other materials involved. It provides teaching and learning materials that deal with familiar chemistry in an unfamiliar context. It also helps to show how the chemical sciences play a part in many unexpected areas of life.

As well as these Teacher's notes, there are five pieces of material:

*The Mary Rose – a historical introduction***.** This is a brief introduction to the historical context of the Mary Rose, her sinking, preservation on the sea bed, subsequent raising and conservation. It does not deal with chemistry in any detail and there are no questions. It can be used by teachers as a background to work on the Mary Rose (found in the materials below) and is suitable for independent reading by students between the ages of 11–19. Students could read it before tackling any of the other sections of the material. Students and teachers could also visit **http://www.maryrose.org** (accessed May 2004) to see more material on the context of the Mary Rose.

Why does the Mary Rose need to be conserved? This material is suitable for 11–16 year-old students. It deals with some of the different materials found in the Mary Rose and how they decay. It is set out as reading material interspersed with questions that relate the chemistry that the students know to the decay (and preservation) of materials found in the Mary Rose.

Preserving the wood from the Mary Rose (1). This is also suitable for 11–16 year-old students. It looks at the chemistry of the preservation of the wood of the Mary Rose and is similar in style to Why does the Mary Rose need to be conserved?, consisting of reading material interspersed with questions.

Preserving the wood from the Mary Rose (2). This material is aimed at post-16 chemistry students and looks at the decay and preservation of the wood in the Mary Rose with reading matter interspersed with questions. The main topics covered are intermolecular forces and polymers.

Metals from the Mary Rose. This is another example of reading matter with questions for post-16 students. It deals with the decay of various metals and the way in which metal objects can be preserved. The main chemical topics are redox reactions and electrochemical cells.

Some material linked to the Mary Rose and suitable for primary students is available separately at **http://www.chemsoc.org/learnnet/**.

Sample material presented is not directly related to the curriculum, it could be used for revision, homework or in case of teacher absence (planned or unplanned). The material can be tackled by students without teacher input provided that students read *The Mary Rose – a historical introduction* first to set the scene.

All the material presented here is available for free download as Word documents from **http://www.chemsoc.org/learnnet/conservation.htm**. This means that teachers may edit them to tailor them to their own requirements. Some teachers may prefer to re-present the reading material as comprehension exercises with all the questions at the end rather than interspersed throughout the material.

The material is also available at the same URL as pdf files that can be read using Adobe Acrobat Reader, available to download from **http://www.adobe.com/products/acrobat/readstep2.html**.

Notes for the teacher on the material

Why does the Mary Rose need to be conserved

At this level, it was not thought necessary or desirable to distinguish between hemicelluloses and celluloses. For the information of the teacher, celluloses are polysaccharides made up of several thousand glucose (sugar) molecules linked together via oxygen atoms (see Figure 1). Hydrogen bonding between the chains lines up the molecules to form bundles called fibrils.

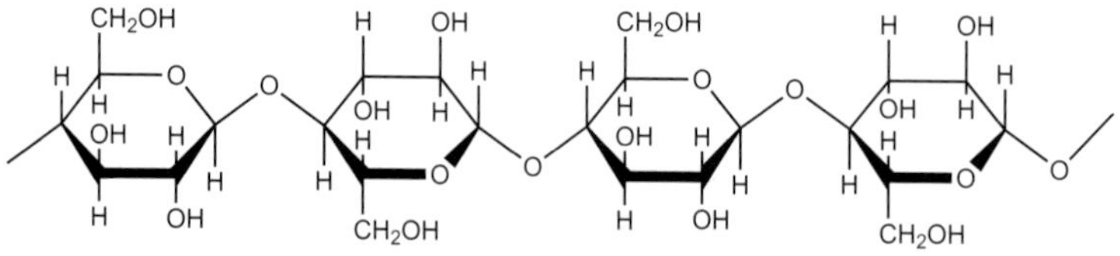

Figure 1 Part of a cellulose molecule

Hemicellulose is also a polysaccharide but has shorter chain than cellulose (between 100 and 200 sugar units) and the chains are branched. In contrast to cellulose, a number of different sugars are found in hemicellulose.

Preserving the wood from the Mary Rose (1)

The point made in the student's material about the chosen conservation method having to allow public viewing was important for financial reasons. The Mary Rose is a major tourist attraction and visits by the public raise a good deal of money which can be spent on conservation. Teachers may wish to stress this economic dimension in the decision-making as well as scientific and purely conservation considerations.

There are ethical debates within the world of museum conservation between conservation (literally 'stopping the rot', but going no further) and restoration (returning an object to how it might have looked originally). Conservators generally try to make sure that any treatment they carry out can be reversed if necessary. There is more

RS•C

material about these issues in the accompanying material on conservation of stone and of plastics.

Preserving the wood from the Mary Rose (2)
It may sound odd to say that in order to preserve the wood from the Mary Rose, it had to be sprayed with water for several years! The long term aim is to have the wood dry but in order to achieve this, spraying with first water, then PEG 200, then PEG 2000 is required so that the drying process is extremely gradual.

Metals from the Mary Rose
Some of the considerations as to which conservation method to use for which metal and in which situation are quite technical. More detail can be found in Mark Jones (ed) *For Future Generations conservation of a Tudor maritime collection*, Portsmouth: The Mary Rose Trust, 2003. This is available from The Mary Rose Trust.

Answers to questions

Answers for Why does the Mary Rose need to be conserved

1. Salt (mostly sodium chloride).
2. (a) Possible answers include precious metals (*eg* gold), some types of stone, glass *etc*.

   (b) Possible answers include more reactive metals, some types of fabrics, paper *etc*.
3. (a) It prevents water coming into contact with the rope.

   (b) Synthetic materials are not attacked by bacteria and other microorganisms.
4. Fish are able to breathe underwater (using their gills to extract the dissolved oxygen).
5. Organic: wood, leather, wool, silk, bone, hemp, horn, flesh
   Inorganic: bone, bronze, copper, cast iron, lead, pewter, gold, glass, stone, pottery, brick, silver, steel.

NB Bones could be placed in either list – organic because it was once-living, inorganic because it is not (mainly) carbon-based. This could be discussed with a suitable group of students.

6. Mainly synthetic (man-made materials, especially synthetic polymers including poly(ethene) (polythene), Nylon™, polyesters such as Terylene™. A significant amount of aluminium would also be expected. Aluminium was not isolated until much later than Tudor times (1827).
7. Temperature, concentration of reactants (pressure of gaseous reactants), surface area of solid reactants, catalysts and (in some cases) light.
8. Metals vary in their intrinsic reactivities (essentially the ease with which they lose electrons). Some metals (aluminium is the classic example) form a protective oxide layer on their surface that prevents further oxidation.
9. The tar prevented water, air and microorganisms coming into contact with the ropes.
10. It already has a high percentage of oxygen in its structure.
11. (a) The (pure) metals known in Tudor times were: iron, copper, zinc, silver, gold, mercury, tin and lead. These are all relatively unreactive metals which means that they could be found uncombined (*eg* gold) or could be fairly easily

RS•C

extracted from their compounds (*eg* by reduction of the oxide with carbon, as in the case of iron).

(b) Gold, it is the least reactive.

12. Gold is an extremely unreactive metal – it is almost unaffected by water or air.

13. Iron rusts quite rapidly on exposure to air and water (especially salt water).

Answers for Preserving the wood from the Mary Rose (1)

1. (a) 6

   (b) A glucose molecule contains six carbon atoms and these all end up as carbon dioxide.

2. (a) Respiration, *ie* the arrow from 'carbon in green plants' to 'carbon dioxide in the
   atmosphere'.

   (b) Photosynthesis, *ie* the arrow from 'carbon dioxide in the atmosphere' to 'carbon in green plants'.

3. As carbon dioxide.

4. Photosynthesis.

5. It would be difficult for the public to view. An enormous tank would be required holding a huge volume of water. It would be impossible for researchers to work under water.

6. The metabolisms of living things slow down at low temperatures.

7. It is less dense than steel (4.5 g cm^{-3} compared with 7.9 g cm^{-3}) and also stronger than steel. This means that smaller and therefore less visually-intrusive supports can be used. It is also less readily corroded than many types of steel.

8. (a) 44

   (b) (i) 194

   (ii) 1998

   (c) The are the approximate relative molecular masses of the two polymers to the nearest round number.

9. See Figure 8. The monomer units are marked with ovals.

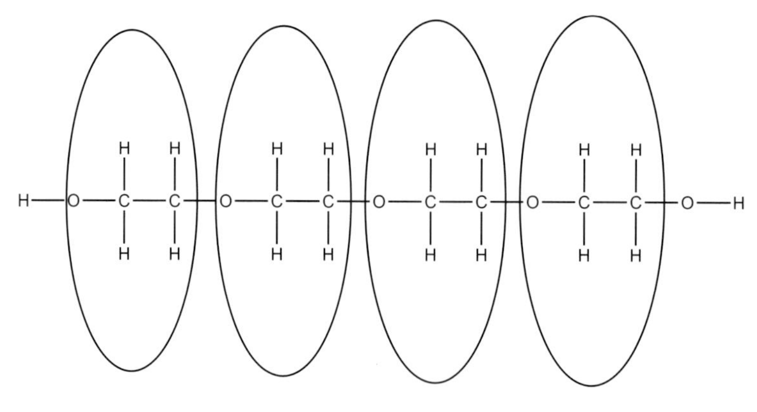

Figure 8

Answers for Preserving the wood from the Mary Rose (2)

1. There are many types of wood from different types of tree and even wood from two different trees of the same type will be different due to conditions of growth *etc.*

2. (a) –OH (alcohol) and R–O–R (ether)

   (b) –OH (alcohol) and R–O–R (ether)

   (c) –OH (alcohol), $-OCH_3$ (ether), –CHO (aldehyde), aromatic rings.

3. It would be difficult for visitors to view it, it would be difficult to work on, a vast mass of water would be required.

4. It is less dense than steel (4.5 g cm^{-3} compared with 7.9 g cm^{-3}) and also stronger. This means that smaller and therefore less visually-intrusive supports can be used. It is less readily corroded than steel.

5. Their metabolisms will slow down at low temperatures.

6. (a) M_r = 200. M_r of the end groups (H and OH) totals 18 which means that the M_r of the repeat unit $(OCH_2CH_2)_n$ must total 200 – 18 = 182. M_r of each repeat unit is 44 and 182/44 = 4.13. So n = 4 to the nearest whole number. It is

   not exactly 4 because the bulk polymer will consist of a mixture of chain lengths and also because the value of 200 has been rounded off to a round number

   (b) M_r = 2000. M_r of the end groups (H and OH) totals 18 which means that the M_r of the repeat unit $(OCH_2CH_2)_n$ must total 4000 - 18 = 1982. M_r of each repeat unit is 44 and 1982/44 = 45.04. So n = 45 to the nearest whole number. It is not exactly 45 because the bulk polymer will consist of a mixture of chain lengths and also because the figure of 2000 has been rounded to the nearest round number.

7. (a) 60°

   (b) The C–O and C–C bond lengths are approximately the same so the ring is an equilateral triangle.

   (c) Approx. 109.5°.

   (d) The four pairs of electrons in the outer shell of the carbon atom mutually repel so they site themselves as far away from each other as possible. This gives an angle of 109.5°.

   (e) The shape of the ring forces the O–C–C angle to be much smaller than the ideal angle. Thus the ring tends to 'spring apart'.

8. Between an electronegative atom (N, O, F) and a hydrogen atom covalently bonded to an electronegative atom.

9. The oxygen atom.

10. See Figure 9

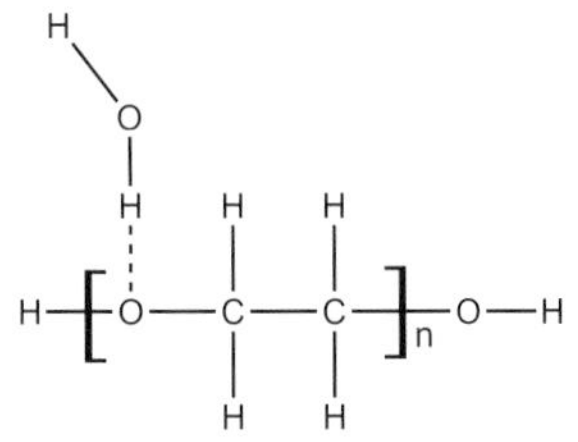

Figure 9 A hydrogen bond between water and PEG

11. (i) –OH

 (ii) –OH, $-OCH_3$, –CHO.

12. (a) There are many possibilities for hydrogen bonding between the polymer and water molecules.

 (b) Larger molecules have higher melting points due to there being stronger intermolecular forces between the molecules.

13. Water.

14. $4OCH_2CH_2 + H_2O \rightarrow H[OCH_2CH_2]_4OH$

Answers for Metals from the Mary Rose

1. About 90, depending on which of the semi-metals are counted.

2. (a) Various answers are possible and should be given credit. Among the most likely responses are aluminium, titanium, cobalt, nickel, tungsten, chromium and vanadium. Some will be found in alloys rather than as pure metals.

 (b) Reasons will depend on the metals suggested and any sensible suggestions should be given credit. Aluminium is used because of its low density, for example.

3. Mercury – it is a liquid at room temperature.

4. $Zn^{+2}O^{-2}(s) + C^{0}(s) \rightarrow Zn^{0}(l) + C^{+2}O^{-2}(g)$

5. $Fe_2O_3(s) + 3C(s) \rightarrow 2Fe(l) + 3CO(g)$
 (an alternative equation with carbon dioxide as the product would be acceptable).

6. Charcoal.

7. The upper equation must be multiplied by 2. This gives the overall equation as

 $2Fe(s) + O_2(aq) + 2H_2O(l) \rightarrow 2Fe^{2+}(aq) + 4OH^{-}(aq)$

8. $H^{+}(aq)$ ions from the dissociation of water would be attracted to the cathode and discharged:
 $2\ H^{+}(aq) + 2e^{-} \rightarrow H_2(g)$

9. Cl^{-} ions will be leached into the water and will form (along with positive ions) a conducting solution that will promote further corrosion.

10. Test a sample of the water with silver nitrate solution. If Cl^{-} ions are present, a white precipitate of silver chloride will be formed:
 $AgNO_3(aq) + Cl^{-}(aq) \rightarrow AgCl(s) + NO_3^{-}(aq)$

11. Chlorine

 $2Cl^{-}(aq)\ -2e^{-} \rightarrow Cl_2(g)$

RS•C

12. See Figure 4.

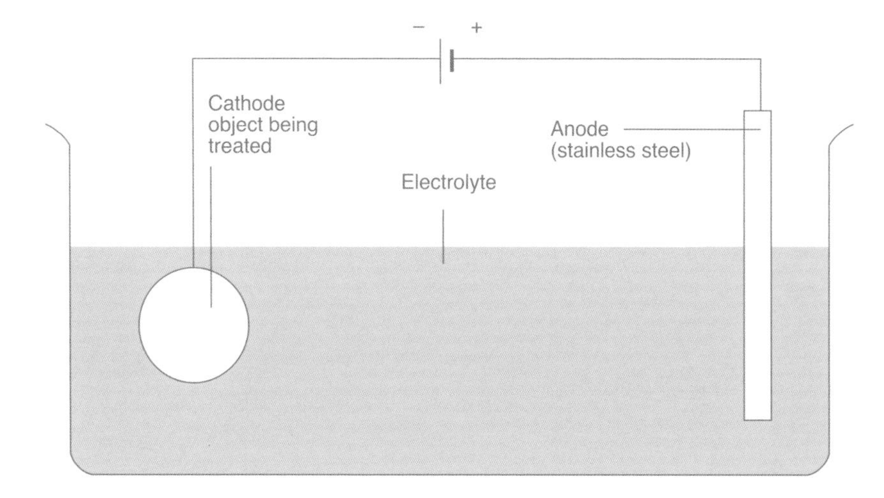

Figure 4 A possible set up for the electrolytic removal of chloride ions from a metal object

13. (a) The Haber process.

 (b) It is unreactive.

 (c) Hydrogen reacts explosively with oxygen in the air.

14. It takes place a high temperature and involves a gas.

15. Some or all of it would be reduced back to iron:
 $Fe_2O_3(s) + 3H_2(g) \rightarrow 2Fe(s) + 3H_2O(l)$

More information

If you have access to the internet, you can visit **http://www.maryrose.org** (accessed August 2004) for a variety of extra information about the Mary Rose.

This site also gives contact details for The Mary Rose Trust from which a number of publications can be obtained. One particularly useful one is *The Mary Rose Museum and Ship Hall*, Portsmouth: The Mary Rose Trust, 2002. It costs just £2.50 and is suitable for all age groups.

There is an article 'The Mary Rose' in *Catalyst* magazine: Chris Young, *Catalyst*, 1998, 8 (3), 1. This is written for 14–16 year old students but could be used with both older and younger children.

A wealth of technical information about the conservation of the Mary Rose can be found in Mark Jones (ed) *For Future Generations conservation of a Tudor maritime collection*, Portsmouth: The Mary Rose Trust, 2003. This is an academic book giving a good deal of technical detail.

RS•C

The Mary Rose – a historical introduction

The Mary Rose is an almost 500 year-old Tudor warship. She was sunk off Portsmouth in a battle with the French in 1545. After spending over 400 years on the sea bed protected from decay by a deposit of silt, her hull was raised in 1982 and is now being preserved in a former dry dock in Portsmouth. Over 19 000 objects were found in and around the ship and these form a sort of 'time capsule' giving historians a unique insight into life in Tudor times. Many scientists, including chemists, are involved in devising and carrying out methods for preserving the ship and her contents from further deterioration so that historians can carry out research and the general public can view this Tudor marvel.

Figure 1 The Mary Rose as she looked in her prime

The Mary Rose was built in Portsmouth and launched in 1511. She was the pride of the Navy in the time of Henry VIII, who ruled from 1509 to 1547. In July 1545 she was sunk during a battle with a French fleet that was attempting to invade England. She went down only about two kilometres (just over a mile) from Portsmouth. We know this because the loss of the ship was recorded at the time – it was watched by a large crowd including the king himself. Nobody is quite sure of the reason she sank. The English version was that the ship was overloaded and mishandled. The French version is that she was hit by French cannon fire, though no French cannon balls have been found inside the wreck.

There was a heavy loss of life when the ship sank. Although the ship was close to shore, nets had been placed over the deck to prevent the enemy boarding her and these stopped many of the crew from escaping as the ship went down.

Why is the Mary Rose so important?

The working lives of wooden Tudor ships were only a few tens of years. Their hulls were attacked by a variety of wood-boring organisms and after this length of time they became uneconomic to repair. So, there are very few existing examples of ships from this era. The Mary Rose is very special because she was a working ship when she sank. Studying the Mary Rose and the objects found with her gives historians a sort of 'snapshot' of life in the mid-1500s because as well as items connected with seafaring and naval warfare, many of these objects relate to normal everyday life at the time. These range from medical instruments to a backgammon set and carpenter's tools to clothing – a small selection is shown in Figure 2.

Figure 2 A small selection of objects found in and around the Mary Rose

How did the Mary Rose survive over 400 years on the sea bed?

What is so unusual about the Mary Rose is that she did not completely rot away during her over 400 years on the sea bed. This was because much of the wreck became covered in silt which protected it (Figure 3).

- She sank onto a bed of silt (a fine mud), and tipped over to the right (starboard).
- A lot of her contents fell into the lowest part of the ship.
- She was anchored to where she fell partly by the weight of her cannons.
- The strong tide washed the silt into the hull where it was trapped and the bottom half of the hull with all its contents was buried there.

RS•C

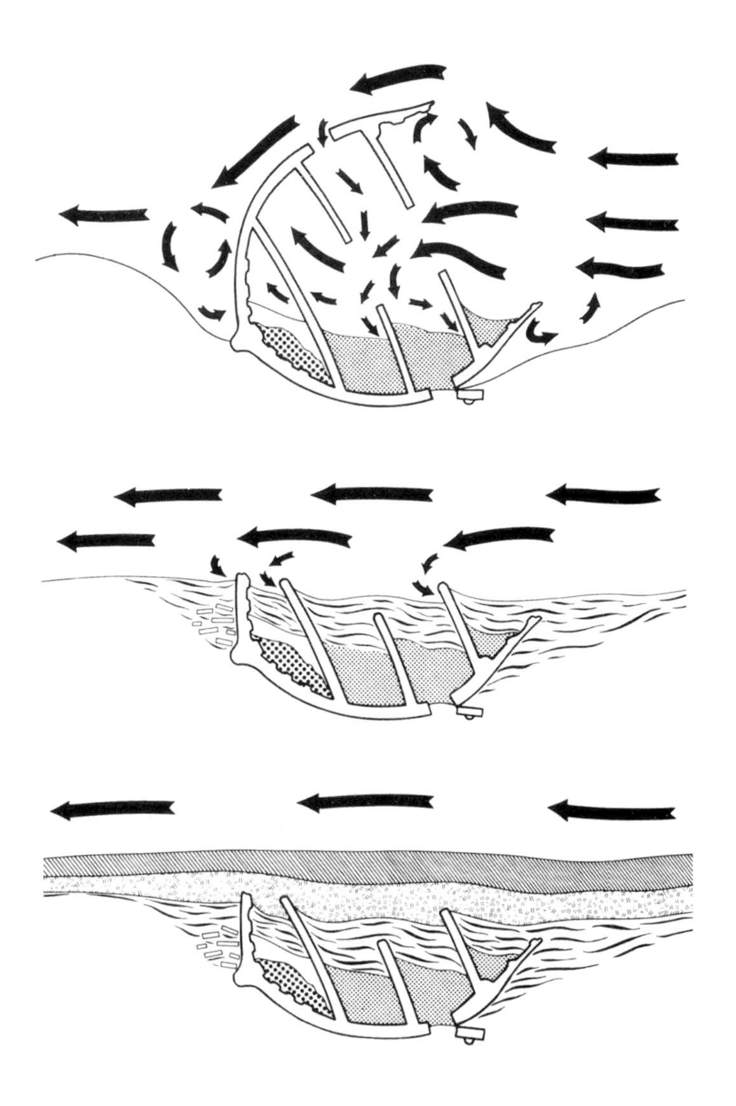

Figure 3 Why the Mary Rose did not rot away
Top: After she sank, the lower part of the Mary Rose's hull filled with silt. This protected the lower timbers but the exposed upper ones were eroded and attacked by a variety of organisms.
Middle: After some time the weakened upper timbers collapsed.
Bottom: Much later a hard layer of clay and crushed shells sealed away the ship and its contents.

The silt acted to protect the part of the ship that was covered by it. It kept the oxygen that is dissolved in seawater away from the ship. In the parts of the ship that were covered, microorganisms, such as fungi and bacteria and larger organisms such as worms could not attack the wood because these all need oxygen. The part of the ship sticking out of the silt rotted away, but the buried section lay preserved under a layer of silt, clay and crushed shells 12 metres (40 feet) below the surface of the sea. Many of the contents of the ship that had fallen into the hull were also preserved in the airless conditions under the silt. As well as the cannons and other weapons you would expect in a fighting ship there were many everyday objects.

RS•C

Raising the Mary Rose

Raising the Mary Rose was a major engineering task because the hull was deep under the water. It was also more than 400 years old and likely to be quite fragile especially once it was out of the water that helped to support its weight.

The problem was tackled by building a steel lifting frame and suspending the hull of the boat from this. Steel cables were used to attach the wood to the lifting frame. Then, the hull was moved (still under water) to a steel support cradle, lined with air bags. The cradle and hull were then lifted to the surface and out of the water. This whole operation took over nine months. It is shown in Figures 4 and 5

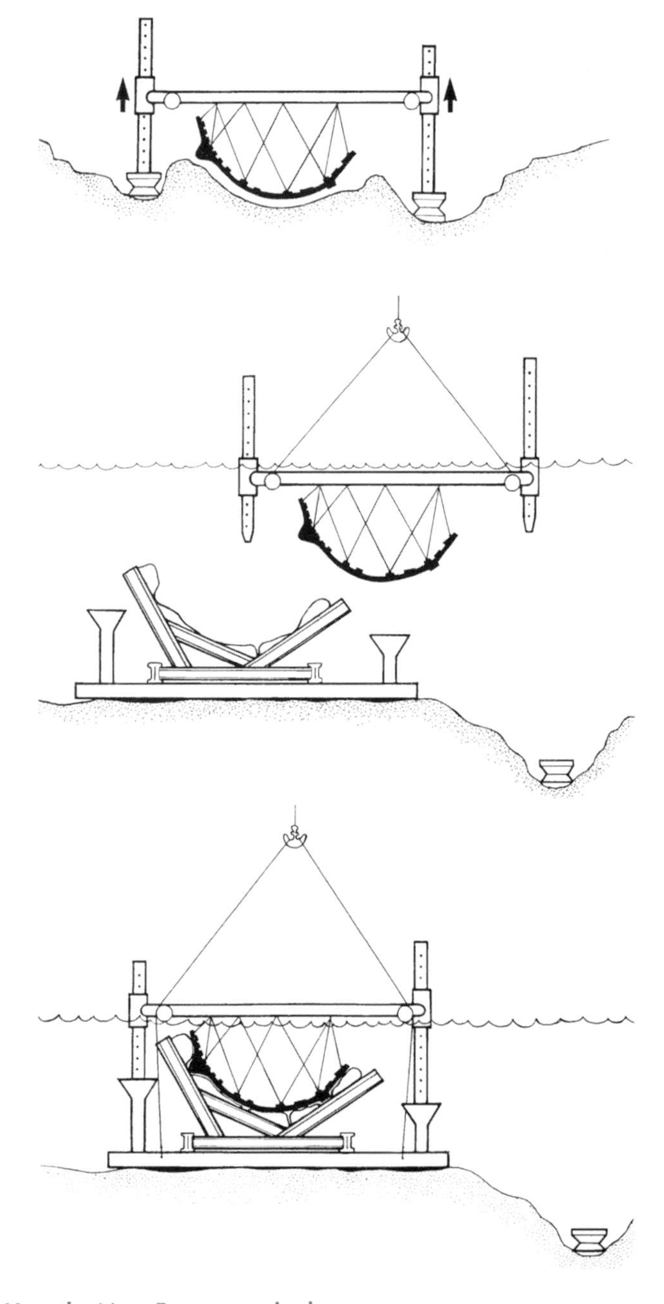

Figure 4 How the Mary Rose was raised
Top The Hull was attached by cables to a steel lifting frame.
Middle Still underwater, the frame and hull were lifted by crane onto a support cradle. The water helped to support the weight of the hull.
Bottom The support cradle holding the hull was lifted out of the water.

RS•C

Figure 5 The Mary Rose sees the light of day after 437 years

Why does the Mary Rose need to be conserved?

The Mary Rose was rediscovered in 1836 and divers brought to the surface some objects including guns. In the 1960s divers carried out surveys of the wreck site and in the 1970s it was decided to try to raise the remains of the ship. In 1982 she was raised and transported to a former dry dock in Portsmouth.

The remains of the Mary Rose and many of the thousands of objects found in and around her are now in the course of being preserved so that historians can find out about life in Tudor times and so that the general public can visit and marvel at her (Figure 6). However, a great deal of effort has to be put in to ensure that, having survived over 400 years on the sea bed, the remains of the Mary Rose and the objects found in and around her can be preserved for posterity. This effort involves many scientists, including chemists, who need to understand the processes that cause decay and devise ways to control them.

RS•C

Figure 6 The Mary Rose as she is today (2004), being sprayed with a waxy solution as part of the preservation process

As well as the ship itself, over 19 000 objects were recovered from the wreck. These range from guns to medical instruments and clothing to coins and include a large range of materials – metals, cloth, rope, leather and many others. During their over 400 years on the seabed, each material decayed in a different way and at a different rate. Now they have been recovered they require different methods of conservation to prevent further decay now they are exposed to air and higher temperatures and to enable them to be studied and/or put on display.

More information

If you have access to the internet, you can visit **http://www.maryrose.org** (accessed ???? 2004) for a variety of extra information about the Mary Rose.

This site also gives contact details for The Mary Rose Trust from which a number of publications can be obtained. One particularly useful one is *The Mary Rose Museum and Ship Hall, Portsmouth*: The Mary Rose Trust, 2002. It costs just £2.50 and is suitable for all age groups.

There is an article 'The Mary Rose' in Catalyst magazine: Chris Young, *Catalyst*, 1998, 8 (3), 1. This is written for 14–16 year old students but could be read by both older and younger children.

RS•C

Why does the Mary Rose need to be conserved?

As well as the ship itself, over 19 000 objects were recovered from the wreck of the Mary Rose. These range from guns to medical instruments and from clothing to coins and include a large range of materials – metals, cloth, rope, leather and many others. During its over 400 years on the seabed, each material decayed in a different way and at a different rate. Now they have been recovered they require different methods of conservation to prevent further decay and to enable them to be studied and/or put on display.

The objects are made from a variety of materials but the main one in terms of sheer bulk is wood. Wood does not last for ever in contact with water and the air that is dissolved in it; it is attacked by bacteria, fungi and wood-boring animals and slowly rots away. Because there are many organisms that attack wood and other similar materials, most archaeological finds are ceramics (*eg* pottery), stoneware or precious metals, like gold. These are not attacked chemically by the oxygen in air, and bacteria and fungi are not able to digest them.

Q1. What is present in seawater as well as water and dissolved air?

Q2. (a) Suggest three materials that you could put in seawater that might last 500 years or more.

(b) Suggest three materials that you could put in seawater that would rot within a year.

Stopping the rot

Materials that would otherwise rot in water can be preserved for longer if their surfaces are coated with a waterproof protective film. In Tudor times, shipbuilders used to use tar or pitch. This is a sticky, waterproof material that is obtained by making cuts in a pine tree and collecting the sticky substance that the tree produces to seal the cuts. It can be melted down (an oven was found on the Mary Rose for this purpose) and the liquefied pitch applied to sails, ropes etc. The ropes aboard the Mary Rose were made from a plant-based material called hemp. They still smell tarry.

Q3. (a) Suggest how applying tar might help to preserve a rope made from hemp.

(b) Modern rope is made from synthetic materials such as Nylon or poly(propene) (polypropylene). Suggest why there is no need to protect these ropes with tar.

Because wood rots relatively easily we might have expected the Mary Rose to have quietly rotted on the seabed. But, something different happened to her.

- She fell onto a bed of silt (a fine mud), and tipped over to the right (starboard).
- A lot of her contents fell into the lowest part of the ship.
- She was anchored to where she fell partly by the weight of her cannons.
- The strong tide washed the silt into the hull where it was trapped and the bottom half of the hull with all its contents was buried there.

Figure 1 shows what is thought to have happened.

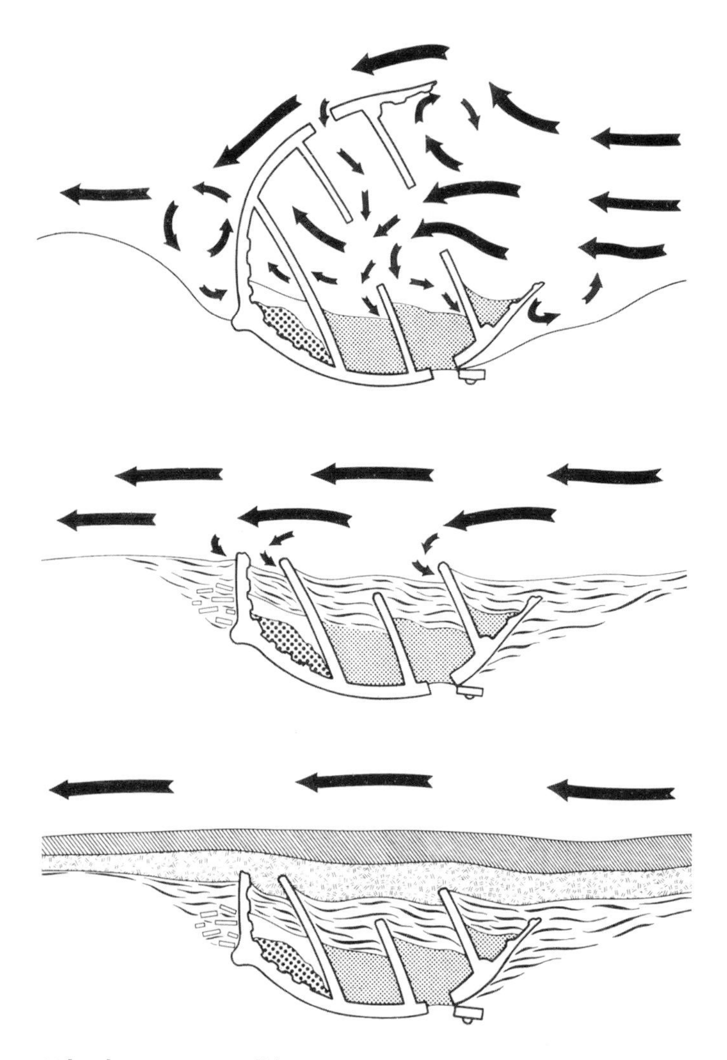

Figure 1 Why the Mary Rose did not rot away

Top After she sank, the lower part of the Mary Rose's hull filled with silt. This protected the lower timbers but the exposed upper ones were eroded and attacked by a variety of organisms.
Middle After some time the weakened upper timbers collapsed.
Bottom Much later a hard layer of clay and crushed shells sealed away the ship and its contents.

The silt acted to protect the part of the ship that was covered by it. It kept the oxygen that is dissolved in seawater away from the ship. In the parts of the ship that were covered, microorganisms, such as fungi and aerobic bacteria and larger organisms such as wood-boring animals (all of which require oxygen) could not attack the wood. The part of the ship sticking out of the silt rotted away, but the buried section lay preserved under a layer of silt, clay and crushed shells 12 metres (40 feet) below the surface of the sea. Many of the contents of the ship that had fallen into the hull were also preserved. As well as the cannons and other weapons you would expect in a fighting ship there were many everyday objects. More than 19 000 separate items were found in the hull. Some are shown in Figure 2.

Q4. Suggest an observation that shows us that seawater contains dissolved oxygen.

Figure 2 Just a few of the items found in and around the Mary Rose

How do different materials decay?

Decay means rotting away. It is a series of chemical reactions. The rates of decay of different materials vary enormously. If we look at the some of the things that survived on the Mary Rose we can divide them into organic (or carbon-based) objects and inorganic ones. Organic materials are generally made from things that were once alive and include wood, leather, and silk.

- Organic materials tend to be attacked biologically by algae, microorganisms (organisms that can only be seen under a microscope) and macroorganisms (larger organisms).
- Inorganic materials tend to be oxidised by oxygen (dissolved in seawater in this case).
- Metals, too, tend to be oxidised.

Table 1 lists some of the material used in Tudor times.

Found in large quantities in the Mary Rose	Not found in the Mary Rose or found only in very small quantities
Wood	Hemp sails
Leather	Horn
Wool and silk	Flesh
Bones	Steel
Bronze / copper	
Cast iron	
Lead	
Pewter	
Gold	
Glass	
Stone	
Pottery	
Brick	

Table 1 Materials used in Tudor times

Q5. Classify each of the materials in the Table into organic or inorganic.

Q6. What types of materials would you expect to find in a modern ship that were not found on the Mary Rose?

What happened over time with either of these groups depended on the conditions that control all rates of reaction. But some things deteriorated more than others, although they were all under the same chemical conditions. Whether they lasted depended to a large extent on their structure and reactivity.

Q7. What are the factors that control the rate of chemical reactions in general?

Q8. Why do some substances of a similar type (*eg* metals) become oxidised at different rates?

How do different materials decay?

Wood

Wood has a complicated structure. Like all once-living materials it is made up of cells. The cells in wood have walls that are made up of a mixture of polymers, mainly celluloses, and lignin. These are present in different proportions depending on the age and type of wood. This makes wood a composite – a material made up of two (or more) other materials, see Figure 3.

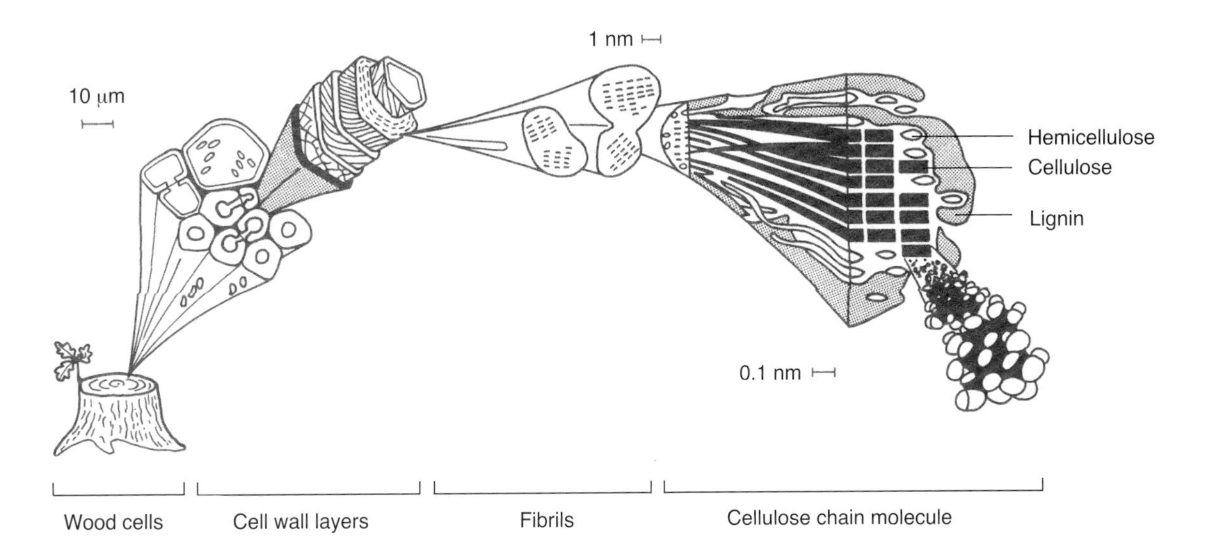

Figure 3 The microscopic structure of wood showing where cellulose and lignin are found. 1 μm is 1 millionth of a metre and 1 nm is 1 thousand-millionth of a metre.

Celluloses (and hemicelluloses) are polysaccharides made up of several thousand sugar molecules linked together. They absorb water, so that the material swells and weakens. After the Mary Rose sank, those materials with a high cellulose content were at first attacked by aerobic bacteria. These will only survive in the presence of oxygen.

Lignin is a polymer with cross-links. This cements the cellulose polymers together. It is a tough material, which does not absorb water and is not easily digested by microorganisms.

Bacteria, fungi and animals such as shipworms all attack wood, see Figure 4.

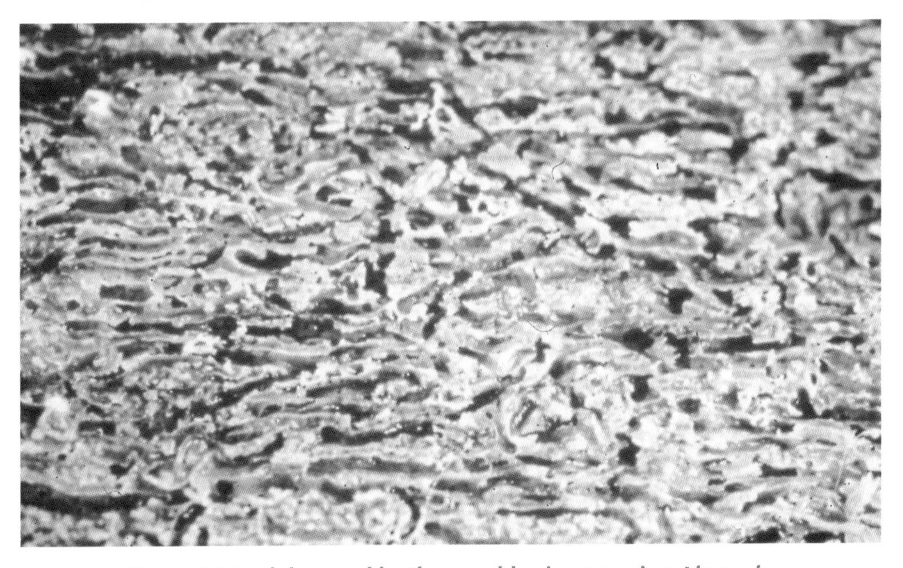

Figure 4 Wood damaged by the wood-boring organism *Limnoria*

Hemp

When the Mary Rose was found, the sails, made of hemp, had almost disappeared. Some of the ropes, also made of hemp, but covered in a tarry, waterproof pitch, were in good condition.

RS•C

Hemp is a plant with a cell structure mostly made of cellulose, with little lignin.

Q9. Explain why most of the sails had disappeared, but the ropes had not.

Ceramics and glass

Ceramics (for example pottery and brick) are heat resistant materials – they will neither melt nor react chemically even at high temperatures. In Tudor times, ceramics would have been moulded from clay and then 'fired' in a hot kiln.

Glasses are usually transparent. Like ceramics they are chemically unreactive but they do soften on heating so that they can be moulded and re-moulded.

Both glass and ceramic materials were in use in Tudor times. They decay very slowly on exposure to oxygen and water and many objects made of these materials have been found in and around the wreck. Chemically, both ceramics and glasses are based on silicon dioxide, SiO_2. Figure 5 shows the arrangement of atoms in a typical glass.

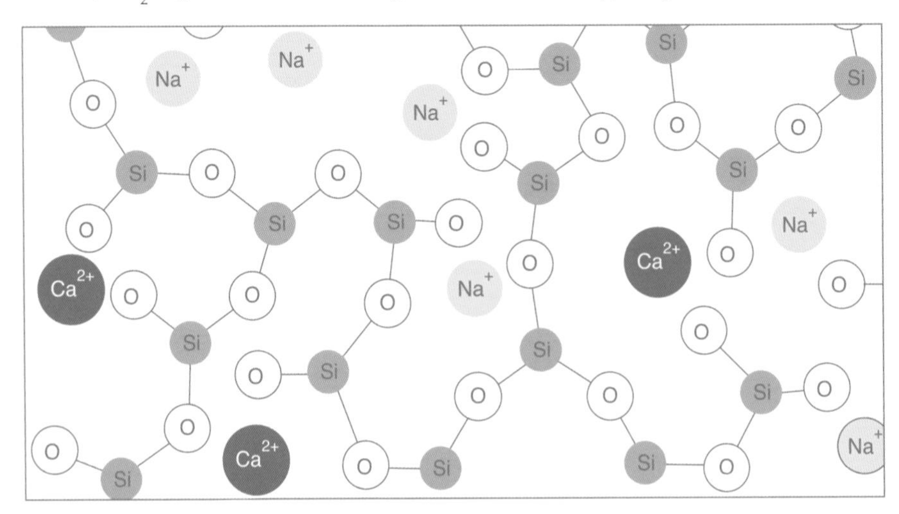

Figure 5 The structure of glass

Q10. Look at the structure of glass. Explain why glass cannot be oxidised.

Metals

Relatively few metals were known and used in Tudor times. Different metals vary widely in how rapidly they react with water and with oxygen. Their reactivities vary from metals like sodium, that react violently when placed in water and tarnish immediately when exposed to air, to gold which will react only with a mixture of concentrated strong acids called aqua regia ('royal water').The list of metals below, is in order of their reactivity, with the most reactive at the top.

Potassium
Sodium
Calcium
Magnesium
Aluminium
Zinc
Iron
Tin
Lead
Copper
Silver
Gold

RS•C

Q11. **(a)** Not all the metals in the list above were known in Tudor times (around 1500). Suggest which ones were known and try to explain. If you have internet access you could use the interactive Periodic Table on the Royal Society of Chemistry website to help you find the elements known at that time (go to **http://www.chemsoc.org/networks/learnnet/periodictable/**, accessed August 2004).

(b) Which of the metals known in Tudor times would be expected to last longest? Explain your answer.

Q12. A gold coin from the Mary Rose can be worth up to £60 000 today. Explain why gold coins were preserved.

Surprisingly, many objects made of iron, such as guns, have survived. Some of them were quickly covered with mussels and other shellfish. When the silt slowly crept over them, the mixture of shells and silt formed a sort of concrete, (called a concretion) which protected the iron. When these objects were brought to the surface, they had to be carefully chipped from the concretion. Under the concretion, the guns were in good condition, see Figure 6.

Figure 6 Chipping concretion from one of the Mary Rose's guns

This method of preservation has not occurred for objects made of other metals – copper, for example. This is because copper compounds are poisonous to most organisms, so the shellfish that formed the basis of the concretion died and did not stick to the object.

Q13. Why is it surprising for objects made of iron to last for over 400 years under the sea?

RS•C

Preserving the wood from the Mary Rose (1)

Although objects recovered from the Mary Rose are made of a wide range of materials (leather, horn, metals, ceramics, glass *etc*), see Figure 1, the most important in terms of sheer quantity is wood.

Figure 1 A selection of articles found in the Mary Rose

How wood decays

Wet wood is attacked by a variety of organisms including wood-boring beetles and worms (collectively called macroorganisms), and bacteria and fungi (collectively called microorganisms). Both these groups of organisms need oxygen from the air – they are called aerobic. These organisms, just like us, produce energy by respiration. The chemical equation for respiration is:

glucose + oxygen → carbon dioxide + water (energy given out)

$C_6H_{12}O_6(s) + 6O_2(g) \rightarrow XCO_2(g) + 6H_2O(l) \quad \Delta H = -2800$ kJ/mol

Q1. The symbol equation for respiration (above) is not completely balanced.

(a) What number does X represent?

(b) Explain your answer.

The organisms obtain glucose from one of the constituents of wood – cellulose – which is a polymer made up of many thousands of glucose molecules linked together.

The main reason that the wood of the Mary Rose had been preserved while it was under the sea was that the silt that covered her had kept out oxygen, producing anaerobic conditions. But once the wet wood was lifted on to dry land and into the air it would become a sitting target for the various organisms to attack. The wood also had to be dried. The scientists were worried that if they left it to dry out in the air, the wood

RS•C

would shrink, become brittle and destroy the hull. There were several conditions that had to be met:

- The wood had to be kept wet because water was needed to prevent shrinkage.
- The wood had to be protected from attack by microorganisms and macroorganisms.
- The method used had to be cheap and safe.
- There had to be access to the hull so that researchers and those people who were looking after the process could work.
- The ship had to be able to be viewed by the public, in order to raise funds needed for research.

Over the ten years or so before the whole ship was lifted, large numbers of timbers had been brought to the surface (about 3000) – some of the timbers that make up the ship are shown in Figure 2.

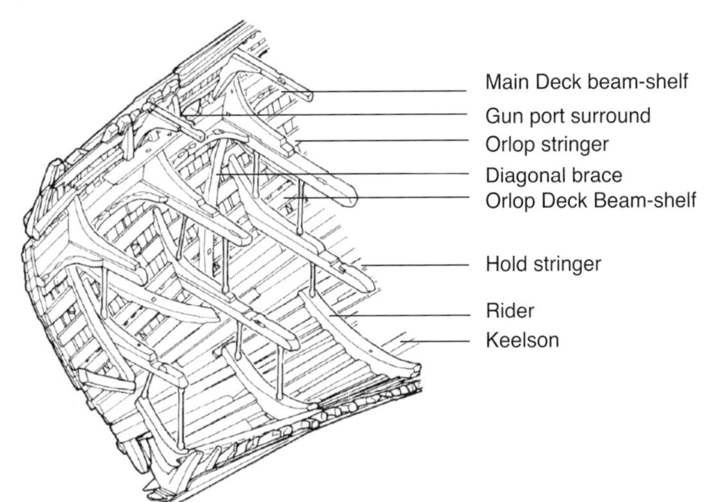

Figure 2 Some of the timber structures that make up the Mary Rose

These timbers, which were made of different types of wood, were studied carefully to learn what needed to be done to stop the wood from decaying fast once it was back on the surface. They were stored underground in the dockyard and the temperature ranged from 9 °C in the winter to 23 °C in the summer. They were studied for attack by organisms at periods.

RS•C

The carbon cycle

Bacteria and fungi use wood as a food (energy) source. This is important part of the carbon cycle. They release the carbon that is locked up in dead wood, and the wood gradually decays away.

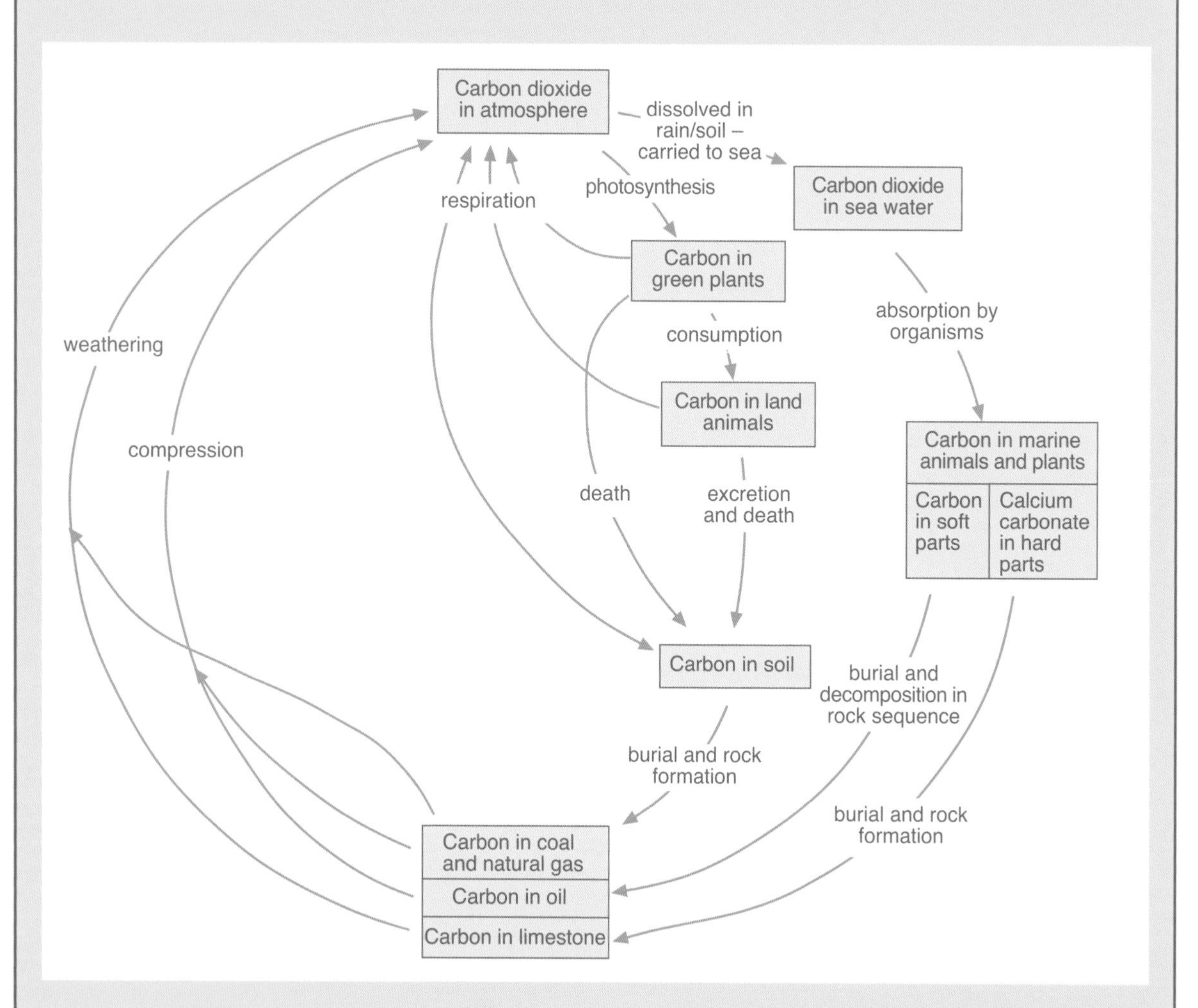

Q2. Identify the parts of the carbon cycle that correspond to

(a) the decay of wood

(b) the formation of wood in the first place.

Q3. In what form is the carbon 'locked up' in the wood recycled to the atmosphere?

Q4. Eventually, the carbon that is recycled to the atmosphere will be 'locked up' again in plant material. What is the name of the process by which this happens?

Bacterial and fungal decay started as soon as the timbers were exposed to the air. A beetle called the wharf-borer (Figure 3) was also found to cause extensive damage. This lays its eggs in old timber and when the eggs hatch out, the larvae burrow through the wood. The scientists found that the greater the attack by fungi, the greater the attack by the wood-borers. They concluded that the beetle larvae were eating the microorganisms.

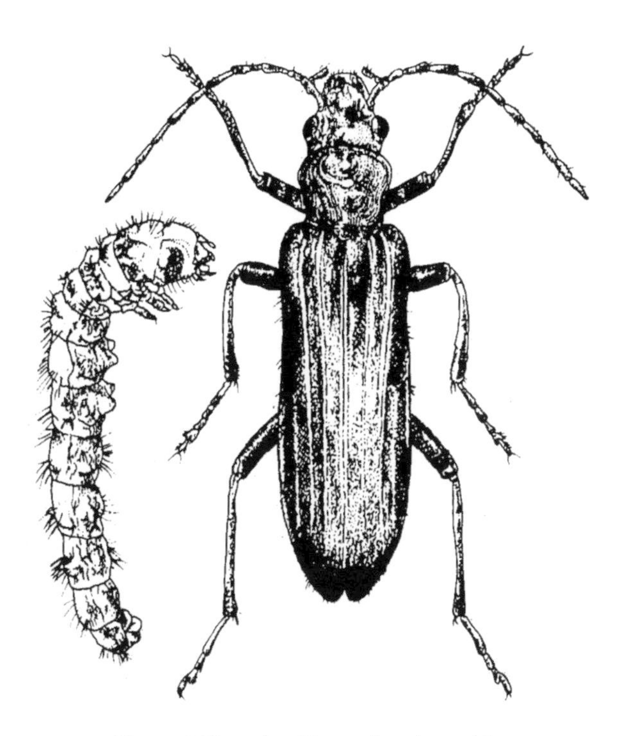

Figure 3 The wharf-borer beetle and larva

They also found that the wharf-borer could get through polythene. So, the scientists now had some information about how to approach the storage of the Mary Rose, once they had brought her out of the water.

The scientists tried some experiments to find out which methods of treatment would work to halt the wood's decay. They found that if they sprayed wood with very cold water, this both kept it wet (stopping shrinkage) and slowed down the growth of the microorganisms and the macroorganisms. They found that they needed a film of water about 1 mm thick on the wood surface.

Q5. Many pieces of wood are preserved by keeping them in a water storage tank. Explain why this was not suitable for the Mary Rose.

Q6. Suggest why low temperatures would slow down the decay caused by living organisms.

The scientists decided on a two-stage process to conserve the hull. The first stage is called a passive stage and involves spraying the wood with cold water. Its function is to give the timber the time it needs to adjust to the conditions of being out of the sea. It is a long process, taking about twelve years. The second stage, called the active stage, involves replacing the water in the wood with a waxy substance to support and strengthen it.

The first stage

Once on the surface, the hull of the Mary Rose was moved to a building once used as a dry dock at Portsmouth. Titanium struts were used to support it. You can see these struts in Figure 4.

RS•C

Figure 4 The Mary Rose being sprayed

Q7. Titanium is a much more expensive metal than steel. Suggest why titanium was used rather than steel for the support struts. You may need to look up some of the properties of the two metals to justify your suggestions.

The old dry dock at Portsmouth was made into a huge refrigerator so that the temperature of the hall could be kept between 2°C and 5°C. The wood was then sprayed with chilled tap water so that the surface of the wood was always covered with a film of water that was 1 mm thick. The water was continually recycled. The water was only turned off for a maximum of four times a day, for an hour each time, so that the boat could be examined, worked on and research carried out. A viewing gallery with glass windows allows the public to see the ship even while the spraying continues, Figure 5. This is important because the admission fees help to fund the conservation work.

RS•C

Figure 5 The viewing gallery

This process continued for 12 years – from 1982 to 1994.

The state of the timber was continually assessed to see how well the method was working.

The second stage

The second stage is called the active stage. It began in 1994 and will come to an end in 2004.The general idea is to very gradually replace the water in the wood with a wax, which will both preserve the wood and support its structure.

This can be done by soaking the wood in a bath of a suitable chemical or by spraying it. Spraying was again chosen for the Mary Rose, because this meant the ship could continue to be seen by visitors, and scientists could still do research.

Polyethylene glycols

The chemical used is called polyethylene glycol 200 or PEG 200. It is a colourless liquid that mixes with water, Figure 6. The ship is sprayed with a mixture of PEG 200 and water; the PEG 200 is gradually absorbed by the wet wood and the water in the wood is slowly replaced by PEG 200.

This will continue until the water in the wood is almost completely gone.

From 2004 a different grade of PEG will be used with a more waxy property, called PEG 2000, Figure 6. This is a solid that dissolves in water. The mixture of PEG 2000 will be sprayed onto the wood until enough of it is absorbed to support and protect the wood. The changeover from PEG 200 to PEG 2000 is due to place during 2004. The reason for using two different grades of PEG is that PEG 200 is more quickly absorbed than PEG 2000 but gives the wood a somewhat tacky feel.

Figure 6 PEG 200 (left) and PEG 2000 (right)

When the wood has absorbed the wax, the boat will be dried out very gradually in the air.

The chemistry of PEG

Polyethylene glycol is a polymer. Polymers are made up of many smaller molecules called monomers. Monomers join together (polymerise) to make a polymer.

$$\text{monomers} \xrightarrow{\text{polymerisation}} \text{polymer}$$

The monomer for both polyethylene glycol 200 and polyethylene glycol 2000 is epoxyethane, whose formula is OCH_2CH_2. The general formula for polyethylene glycol is $H(OCH_2CH_2)_nOH$. Polyethylene glycol 200 has four monomer units in its chain (plus a H atom at one end and an -OH group at the other) so n = 4, see Figure 7. Polyethylene glycol 2000 has 45 monomer units in its chain (plus end groups) so n = 45.

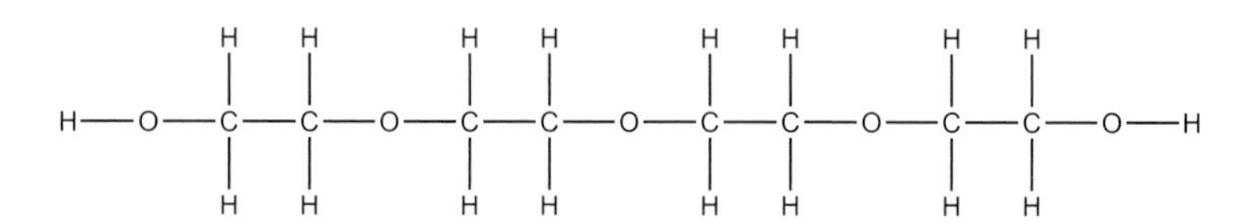

Figure 7 The PEG 200 molecule

Q8. **(a)** What is the relative molecular mass of epoxyethane, OCH_2CH_2? (A_rs: C = 12, O = 16, H= 1)

(b) Use this to help you work out the relative molecular mass of
(i) polyethylene glycol 200
(ii) polyethylene glycol 2000.

(c) Suggest the significance of the 200 and the 2000 in the names PEG 200 and PEG 2000.

Q9. On a copy of the formula of PEG 200, Figure 7, mark the four monomer units.

RS•C

Preserving the wood from the Mary Rose (2)

There were around 19 000 artefacts found in and around the Mary Rose. These were made from a large range of materials – metals, ceramics, glasses, horn, hemp, linen *etc etc*. However, in terms of sheer bulk, the main material that required preservation was wood.

The structure of wood

Wood has a complex structure. Like all once-living materials it is made up of cells. The cells in wood have walls that are made up of a mixture of polymers, mainly celluloses (40–50%), hemicelluloses (20–30%) and lignin (20–30%). These are present in different proportions depending on the age and type of wood. This makes wood a composite – a material made up of two (or more) other materials, see Figure 1.

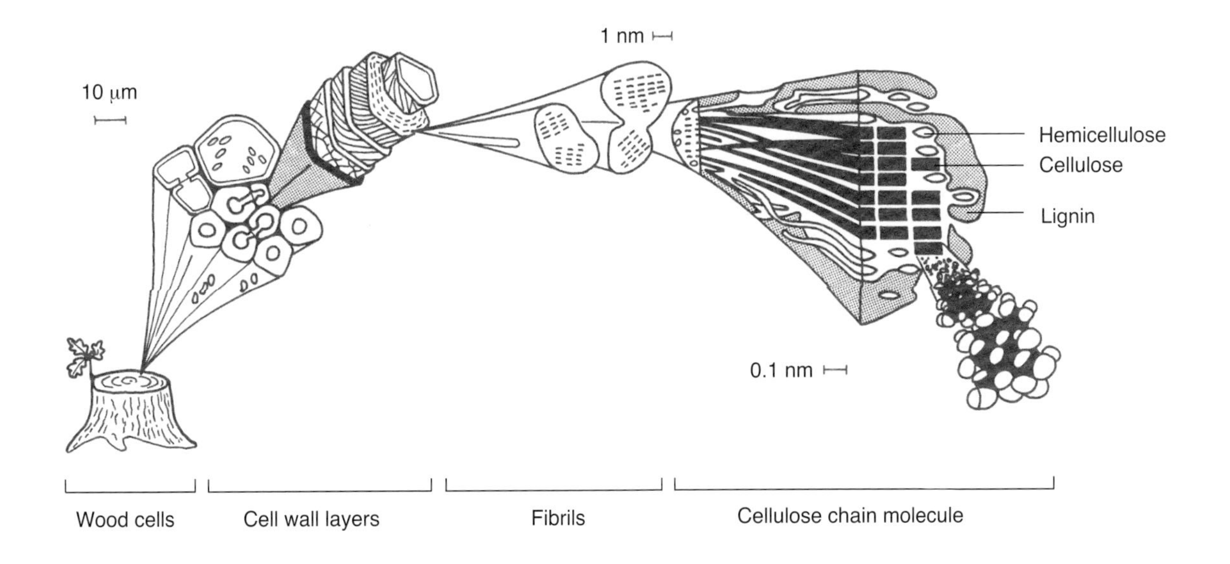

Figure 1 The structure of wood.
1μm is 1 millionth of a metre, 1 nm is one thousand-millionth of a metre

Q1. Why is it an oversimplification to talk about wood as a single material?

Celluloses are polysaccharides made up of several thousand glucose (sugar) molecules linked together via oxygen atoms, see Figure 2. Hydrogen bonding between the chains lines up the molecules to form bundles called fibrils.

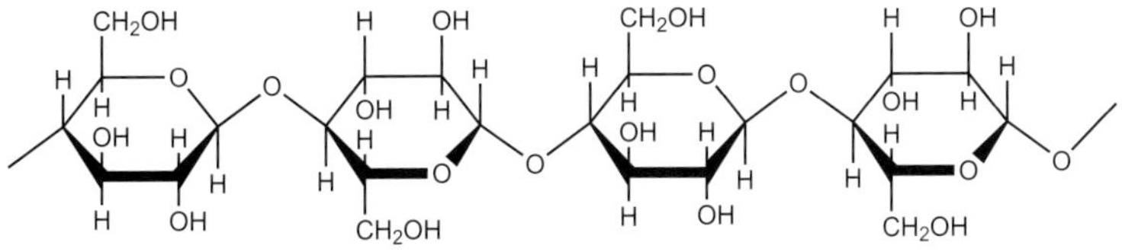

Figure 2 Part of a cellulose molecule

Hemicellulose is also a polysaccharide but has shorter chains than cellulose (between 100 and 200 sugar units) and the chains are branched. In contrast to cellulose, a number of different sugars are found in hemicellulose.

Lignin is a polymer with cross-links, see Figure 3. This cements the cellulose polymers together. It is a tough material, which does not absorb water and is not easily digested by microorganisms.

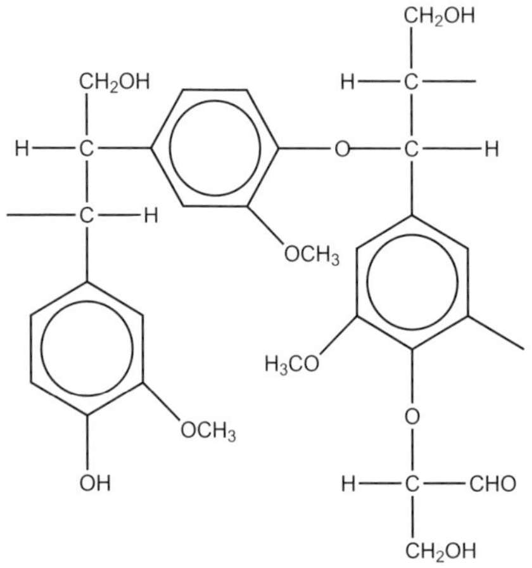

Figure 3 Part of a lignin molecule

Q2. What functional groups are present in a molecule of

(a) cellulose

(b) hemicellulose

(c) lignin?

The decay of wood

When wood is immersed in water, the cellulose and hemicellulose molecules absorb water, so that the wood swells and weakens. They absorb water because hydrogen bonds form between water molecules and the –OH groups of the glucose molecules that make up cellulose and hemicellulose, see below.

Other causes of decay are beetles such as wharf-borers, see Figure 4, and bacteria and fungi. Most of these organisms require oxygen to live and are called aerobic. The reason that the wood of the Mary Rose's hull that was protected by silt rotted very little is that the silt protected it from air, producing an anaerobic environment. The worms and other larger organisms are collectively called macroorganisms while the bacteria and fungi are called microorganisms.

Figure 4 Damage to an oak beam caused by the larva of a wharf-borer

Most of the bacteria and fungi that attack wood need oxygen to survive. Organisms that need oxygen are called aerobic. These organisms, just like us, produce energy by respiration. The chemical equation for respiration is:

glucose + oxygen → carbon dioxide + water (energy given out)

$C_6H_{12}O_6(s) + 6O_2(g) \rightarrow 6CO_2(g) + 6H_2O(l)\ \Delta H = -2800\ kJ\ mol^{-1}$

The only organisms that are found living in oxygen-free silts such as the one that covered Mary Rose are anaerobic bacteria. The rates of decay caused by these organisms are very slow.

Preserving the wood of the Mary Rose

Once the wet wood of the Mary Rose's hull was lifted onto dry land and into the relatively warm air it became a sitting target for both microorganisms and macroorganisms to attack. To preserve it, the wood had to be dried in the long term. The scientists were worried that if it left to dry out in the air, the wood would shrink and destroy the hull. There were several conditions that had to be met:

- The wood had to be kept wet to prevent shrinkage.
- The wood had to be protected from attack by microorganisms and macroorganisms.
- The method used had to be relatively cheap and safe.
- There had to be access to the hull so that researchers and those people who were looking after the process could work.
- The ship had to be able to be viewed by the public, in order to raise funds needed for research.

Before the Mary Rose was lifted from the sea bed, the scientists tried some experiments on smaller pieces of timber to find out what would work. They also consulted the work of other conservators who had tackled similar problems, in particular the Swedish ship the Vasa. They concluded that if they sprayed wood with very cold water, this would keep it wet and slow down the growth of both the microorganisms and the macroorganisms. They found that they needed a film of water about 1 mm thick on the wood surface.

Q3. Many pieces of archaeologically important wood are preserved by keeping them in a water storage tank. Use the list above to say why this was not suitable for the Mary Rose.

It would not be practical to spray the ship with water for ever. So, the conservators decided on a two-stage process. The first stage, spraying with water, is called the passive stage. Its function is to give the timber the time it needs to adjust to the conditions of being off the sea bed. It is a long process, taking about twelve years. The second stage, the active stage, involves displacing the water in the timbers with a waxy substance called polyethylene glycol (PEG for short).

Problems with conservation

One of the problems with the conservation of archaeological relics is that conservators have to be careful about making changes that they cannot reverse. The following (true) story illustrates this.

One of the first objects to be raised to the surface from the wreck of the Swedish ship Vasa was a slab of butter. After 300 years under the sea in cold temperatures it was hard and solid but as soon as it was exposed to everyday temperatures it began to melt and was soon gone.

Conservators need to avoid problems of this sort on a larger scale. In the case of the Mary Rose, any damage caused by drying the hull too quickly would have been impossible to reverse.

The passive stage

Once on the surface, the hull was moved to a former dry dock at Portsmouth. Titanium struts were used to help support it, see Figure 5.

Figure 5 The Mary Rose with the sprayers in action. The titanium struts can be seen clearly

Q4. Titanium is a much more expensive metal than steel. Suggest why titanium was used rather than steel to support the Mary Rose's hull. You will need to look up some of the properties of the two metals to justify your answer.

The dry dock at Portsmouth was made into a huge refrigerator so that the temperature of the hall could be kept between 2° and 5°C.

The wood was then sprayed with chilled tap water so that the surface of the wood was always covered with a film of water that was 1 mm thick. The water was continually recycled. The water was only turned off for a maximum of four times a day, for an hour each time, so that the boat could be examined, worked on and research carried out.

This process continued for 12 years – from 1982 to 1994.

The state of the timber was continually assessed to see how well the method was working.

Q5. Why would very cold water slow down the growth of microorganisms and macroorganisms?

The active stage

This began in 1994 and is still going on today (2004). The general idea is to very gradually replace the water in the wood with a type of wax, which will both preserve the wood and support its structure.

This can be done by soaking the wood in a bath of a suitable chemical or spraying it. Spraying was again chosen for the Mary Rose, because this meant that the ship could continue to be seen by visitors and that sprays could be turned off from time to time to allow scientists to continue their research.

Polyethylene glycols

The chemical chosen is called polyethylene glycol or PEG for short, see Figure 6. PEG is not a single substance – it refers to a range of compounds all of which are polymers of the monomer epoxyethane, see Figure 7.

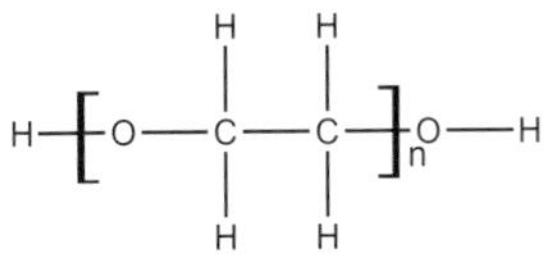

Figure 6 PEG

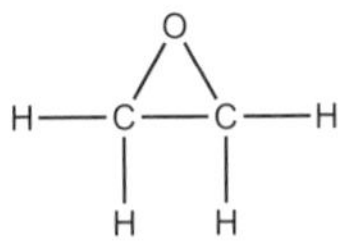

Figure 7 Epoxyethane

Two grades of PEG are used – PEG 200 and PEG 2000, see Figure 8.

RS•C

Figure 8 PEG 200 (left) and PEG 2000 (right)

The numbers refer to the relative molecular masses of the two polymers and therefore are a measure of the number of monomers in each polymer – called the chain length. That is, the two polymers have different values of n. The ship was first sprayed with a solution of PEG 200 in water. This will take place for several years, during which the PEG 200 will displace water from the timbers. This will be followed by spraying for several more years with a solution of PEG 2000, which will eventually displace the PEG 200. The changeover is about to take place at the time of writing (May 2004). When this process is complete, the solid PEG 2000 will support and strengthen the decayed timbers. The reason for using two different grades of PEG is that PEG 200 is more quickly absorbed that PEG 2000 but gives the wood a somewhat tacky feel.

Q6. The general formula of PEG is $H[OCH_2CH_2]_nOH$. Work out the number of monomer units in

(a) polyethylene glycol 200

(b) polyethylene glycol 2000.

Remember to take account of the end groups (the groups outside the brackets) in your calculation.

Hydrogen bonding

PEG mixes with water because hydrogen bonds form between the PEG molecules and water molecules.

Both water and PEG will also form hydrogen bonds with the lignin, hemicellulose and cellulose molecules in wood. This is why they are both absorbed by wood.

PEG 200 is a colourless liquid that mixes with water. The mixture of PEG and water is gradually absorbed by the wet wood and the water is slowly replaced by PEG 200.

Q8. Between what sorts of atoms do hydrogen bonds form?

Q9. What atom in the main chain of a PEG molecule will form a hydrogen bond with a water molecule?

Q10. Draw a diagram to show this hydrogen bond.

Q11. Which groups in (i) cellulose and (ii) lignin will form hydrogen bonds with PEG and with water.

RS•C

Epoxyethane

Epoxyethane is an important industrial chemical. It has no uses as an end product but is an intermediate used for making a range of other chemicals including anti-freeze (ethane-1,2-diol), detergents, polymers and lubricants. 11 million tonnes are made each year world-wide.

It is made from ethene, which itself is made by cracking long chain alkanes obtained from the distillation of crude oil. Ethene is then heated along with oxygen from the air under pressure and using a silver catalyst to produce epoxyethane:

$CH_2{=}CH_2 + {}^{1}/{2}\,O_2 \rightarrow CH_2OCH_2$

Note that epoxyethane actually exists as a ring, see Figure 7.

To make ethane-1,2-diol (ethylene glycol) anti-freeze, epoxyethane is reacted with water:

$CH_2OCH_2 + H_2O \rightarrow CH_2OHCH_2OH$

Epoxyethane is highly reactive, due largely to so-called ring strain in its three-membered ring, see Figure 7.

Q7. **(a)** Look at Figure 6. What is the approximate O–C–C bond angle in epoxyethane?

(b) What assumption are you making in your answer?

(c) What is the normal O–C–C angle in, say, ethanol?

(d) Explain your answer.

(e) Suggest how your answers to (a) and (c) help to explain the reactivity of epoxyethane.

The second phase

The second phase of PEG treatment has just begun (2004). The solution of PEG 200 used to spray the hull has been replaced with one of PEG 2000. PEG 2000 is a waxy solid that dissolves in water. The mixture will be sprayed onto the wood and will displace the PEG 2000 until enough of it is absorbed to support and protect the wood. This process, too, will take several years.

When the wood has absorbed the wax, the boat will be dried out very gradually in the air. Any cavities in the wood that originally contained cellulose will then contain solid PEG 2000.

Q12. Explain why PEG 2000

(a) dissolves in water

(b) is a solid whereas PEG 200 is a liquid.

The chemistry of PEG

Polyethylene glycol is an addition polymer. Polymers are made up of many smaller molecules called monomers. Monomers join together (polymerise) to make a polymer.

$$\text{monomers} \xrightarrow{\text{polymerisation}} \text{polymer}$$

Q13. Look at the end groups (those outside the square brackets in the formula of PEG, Figure 6). The word equation above is a slight oversimplification. What molecule other than epoxyethane is required to form the end groups in the polymerisation process?

Q14. Complete the equation below for the polymerisation of epoxyethane to form PEG 200:

$?\,OCH_2CH_2 + ? \rightarrow H[OCH_2CH_2]_4OH$

RS•C

Metals from the Mary Rose

Around 3000 metal objects were found in and around the Mary Rose. These include a wide selection of items including cannon and shot (*ie* cannon balls), utensils and cutlery, coins, navigational instruments, medical instruments and the ship's bell, see Figure 1. There was also a variety of metals, although not as many as would be found in a modern ship. This is because far fewer metal were known in the early 1500s than today. In fact the only metallic elements known at the time were iron, copper, zinc, silver, tin, gold, mercury and lead.

Figure 1 A selection of objects found in and around the Mary Rose

Q1. Use a Periodic Table to help you find out how many metals are known today.

Q2. **(a)** Suggest metals that would be found in a modern ship that were not known in the time of the Mary Rose.

(b) Suggest reasons for the use of these metals.

Q3. Which of the metals known in Tudor times would not be found in the Mary Rose? Explain your answer.

The metals found in the Mary Rose can be divided into ferrous (wrought iron, cast iron and steels) and non-ferrous (copper and its alloys, lead and its alloys, tin and its alloys, silver and gold).

Extraction of metals

Apart from gold, almost no metals are found uncombined. They are usually found in compounds combined with oxygen, or sometimes sulfur, in which the metal exists as positive ions. Metals are often extracted from their compounds by heating with a reducing agent such as carbon. This removes the oxygen as carbon monoxide. For example

$ZnO(s) + C(s) \rightarrow Zn(l) + CO(g)$

or ionically

$Zn^{2+}O^{2-}(s) + C(s) \rightarrow Zn(l) + CO(g)$

Q4. Put in the oxidation numbers of all the elements in the equation for the extraction of zinc.

Note that this is a reduction process as it involves the gain of electrons by the metal.

This method will only work for relatively unreactive metals, which helps to explain why so few metals were known in Tudor times.

Q5. Write an equation to show the extraction of iron from iron ore (Fe_2O_3) by carbon.

Q6. Suggest a source of carbon that was available in Tudor times.

Corrosion of metals

Corrosion of metals involves metals losing electrons to form positive ions. In this sense it is the opposite of the extraction process described in the box. These electrons must be given to some other species such as another metal. In aqueous conditions such as below the sea, an electrochemical cell is formed between two metals that are in contact, such as iron and copper. The more reactive metal, iron (called the anode), loses its electrons more readily and transfers them to the copper (called the cathode) as a flow of electric current. At the copper cathode the electrons react with oxygen dissolved in the water to form hydroxide ions:

$Fe(s) \rightarrow Fe^{2+}(aq) + 2e^-$

$O_2(aq) + 2H_2O(l) + 4e^- \rightarrow 4OH^-(aq)$

The copper does not corrode.

Q7. Write an overall equation for the corrosion process by adding the two half equations together. You will need to multiply one of the half equations by a suitable factor so that both half equations involve the same number of electrons.

The electrical circuit is completed by the seawater (called the electrolyte) via the movement of sodium and chloride ions.

In the laboratory we could mimic what happens with the setup in Figure 2 which shows the iron and copper as separate strips connected by a wire. It also shows the movement of ions and electrons.

RS•C

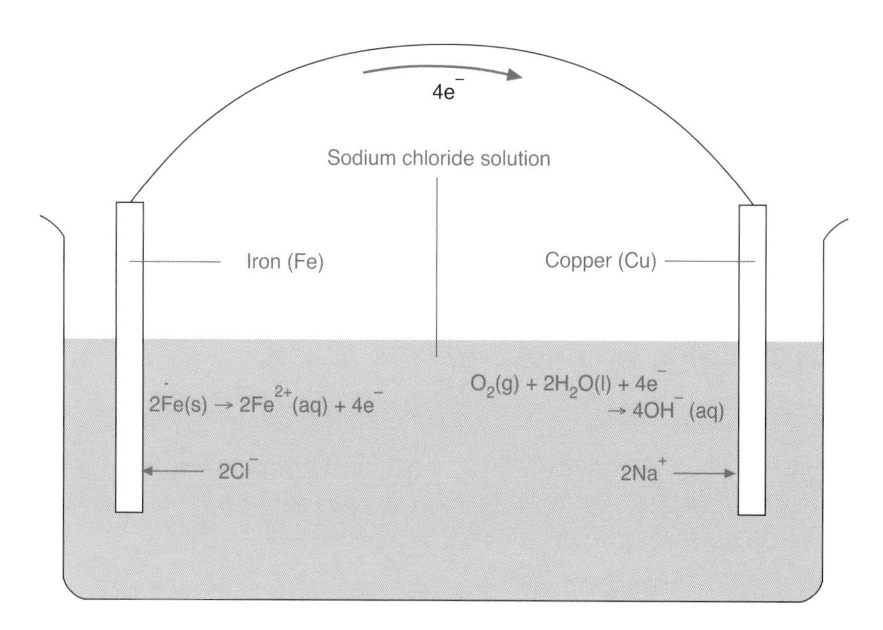

Figure 2 The electrochemical corrosion of iron in the laboratory

At first sight it is difficult to see how a single metal corrodes in this way, but in fact there are always impurities in metals (especially in metals extracted in Tudor times) and cells are set up at the points of contact between the metal and its impurities.

So electrochemical corrosion requires five conditions:

- An anode – the metal that is corroding.
- A cathode – a less-reactive metal that does not corrode.
- Contact between the anode and cathode so that electrons can flow from anode to cathode.
- An electrolyte – this is a solution containing ions that completes the electrical circuit.
- A reactant, such as dissolved oxygen, at the cathode that 'mops up' the electrons released by the anode.

If any one of these is missing, corrosion will not take place. This explains why all the metal objects on the Mary Rose did not completely corrode away during their 400 years under the sea; the silt that covered the ship sealed it away from oxygen. Once the oxygen originally in the sea water was used up no more corrosion could occur.

Concretion

Many of the metal objects from the Mary Rose were found encased in a layer of calcium carbonate – this is called concretion, see Figure 3. Concretion further protected the metal from corrosion.

Figure 3 A cannon ball removed from its protective concretion

Concretion is caused by the OH^- ions arising from the initial electrochemical corrosion of the metal (see above). These react with calcium hydrogencarbonate that is dissolved in sea water forming a layer of insoluble calcium carbonate (this is like limestone) around the metal object which seals it off from further oxygen and stops the corrosion.

$$Ca(HCO_3)_2(aq) + 2OH^-(aq) \rightarrow CaCO_3(s) + H_2O(l) + CO_2(g)$$

Once the initial layer of calcium carbonate is established, marine organisms (such as mussels, limpets and oysters) can colonise the object and their shells, also made of calcium carbonate, can add to the concretion. However, this did not happen with all metals - copper, for example, is toxic to marine organisms.

Conserving metal objects from the Mary Rose

The conditions around the Mary Rose prevented or slowed down corrosion of metal objects in many cases. However, once objects were recovered and exposed to oxygen again, corrosion could resume. Conservators wished to prevent this happening. Their aims were threefold:

- To stop the corrosion process.
- To leave the shape of the object unchanged.
- To leave the metallic structure of the object unchanged.

To do this they had to tackle a problem caused by the long immersion of the artefacts in sea water. This was that during this immersion the surface layers of metal objects had absorbed chloride ions from the sea water. This could cause two main problems.

Firstly, any water that condensed on the surface of an object would tend to dissolve these ions and form an electrically-conducting solution that would form the electrolyte for electrochemical corrosion. This, together with oxygen from the air could form the ideal conditions for corrosion. To avoid this problem, objects would have to be stored and displayed in extremely dry air – difficult to achieve in a museum, for example, where visitors are continually breathing out moist air.

Secondly, under some conditions, the chloride ions could react to form hydrochloric acid which would itself corrode the metal.

For both these reasons it was necessary to remove the chloride ions.

Three methods were considered:

1. Washing the objects with water to dissolve the chloride ions out of the surface layer.
2. An electrolytic method in which the artefact is made the cathode (negative electrode). The negative chloride ions would be attracted to the positive electrode and drawn out of the metal.
3. Heating the object in an atmosphere of hydrogen. The chloride ions react with the hydrogen to form hydrogen chloride gas which is then sucked out of the furnace.

Experience suggested that method 1 could take up to five years to remove all the chloride ions. During this time, the object would be immersed in a solution containing chloride ions and would therefore be corroding.

Method 2 would be quicker, but could still take two years. There were two disadvantages. Firstly an electrical connection would have to be made to the object and this might damage it. Secondly, there is a possibility of hydrogen gas being produced at the cathode. If this were too vigorous, it could cause damage to corrosion already present on the cathode. As with method 1, during treatment the object would be immersed in a solution containing chloride ions and would therefore be corroding.

Q8. Explain with suitable equations how hydrogen could be produced at the cathode.

Q9. If method 1 is used, it is necessary to change the water regularly. Explain why this is needed.

Q10. If method 1 is used, it is necessary to test the water from time to time for chloride ions to confirm when all the chloride had been leached out. Suggest how this test might be done and give any relevant equation(s).

Q11. If method 2 is used, what gas will be formed when the chloride ions reach the anode. Give an equation for the formation of this gas.

Q12. Draw a diagram to show a possible setup for method 2.

For iron objects, heating with hydrogen to 850 °C was the method actually chosen – even this had to be done for a week. A special furnace had to be built to hold objects as big as cannon. The reaction that is taking place may be represented as

$$2FeCl_3(s) + 3H_2(g) \rightarrow 2Fe(s) + 6HCl(g)$$

For safety reasons, the gas used in the furnace was actually a mixture of hydrogen and nitrogen. It turned out to be cheap and convenient to make this gas mixture by decomposing liquid ammonia using a catalyst:

$$2NH_3(l) \rightarrow N_2(g) + 3H_2(g)$$

Q13. **(a)** What industrial process is this the reverse of?

(b) Why does the presence of nitrogen not affect the chemical reactions taking place in the furnace?

(c) Why might pure hydrogen pose a safety risk?

Q14. Give reasons why the reaction with hydrogen was the fastest process.

Q15. A simplified formula for rust is Fe_2O_3. Suggest what would happen to rust when heated with hydrogen. Write an equation for this reaction.

Question 15 highlights one problem with the hydrogen furnace method of treatment – it can actually change the metal and its corrosion products rather than simply preventing further corrosion. For this reason samples of the corroded metal were taken and stored before objects were treated in this way. Conservators are keen that objects are changed as little as possible by the methods they use.

For different metals, one of the other methods of conservation was normally chosen. In general, which metal was treated with which method is shown in the Table. Sometimes the method chosen depend on the particular object as well as the metal it was made from. For example very fragile objects might be damaged by making an electrical connection and this could rule out the electrolysis method.

Metal	Method of conservation	Notes
Iron	Hydrogen furnace	
Copper	Washing with water	Hot water was used as this contains less dissolved oxygen which might have caused more corrosion. This outweighed the increased rate of corrosion caused by the temperature of the water.
Bronze (an alloy of copper and tin)	Electrolysis	
Brass (an alloy of copper and zinc)	Washing in water containing a corrosion inhibitor	Objects were coated with lacquer after soaking
Lead	Washing with hydrochloric acid	The acid dissolved away any concretion and left a film of insoluble lead(II) chloride on the metal surface which protected against further corrosion
Pewter (an alloy of tin and lead)	Electrolysis	Except where the objects were too fragile for an electrical connection to be made.
Silver	Washing or hydrogen reduction	
Gold	None required	Gold is too unreactive to corrode

RS•C

Your notes

Your notes

Your notes